Lecture Notes in Artificial Intelligence   4795

Edited by J. G. Carbonell and J. Siekmann

Subseries of Lecture Notes in Computer Science

T0224428

Frank Schilder   Graham Katz
James Pustejovsky (Eds.)

Annotating, Extracting and Reasoning about Time and Events

International Seminar
Dagstuhl Castle, Germany, April 10-15, 2005
Revised Papers

 Springer

Series Editors

Jaime G. Carbonell, Carnegie Mellon University, Pittsburgh, PA, USA
Jörg Siekmann, University of Saarland, Saarbrücken, Germany

Volume Editors

Frank Schilder
Thomson Corp., R&D
610 Opperman Drive, Eagan, MN 55123, USA
E-mail: frank.schilder@thomson.com

Graham Katz
Georgetown University, Department of Linguistics
37th and O Streets, NW, Washington, DC 20057, USA
E-mail: egk7@georgetown.edu

James Pustejovsky
Brandeis University, Computer Science Department
415 South St., Waltham, MA, 02454, USA
E-mail: jamesp@cs.brandeis.edu

Library of Congress Control Number: 2007937633

CR Subject Classification (1998): I.2.4, I.2.6-7, I.7.2, H.2.8, H.3, F.4

LNCS Sublibrary: SL 7 – Artificial Intelligence

ISSN   0302-9743
ISBN-10   3-540-75988-3 Springer Berlin Heidelberg New York
ISBN-13   978-3-540-75988-1 Springer Berlin Heidelberg New York

Springer is a part of Springer Science+Business Media

springer.com

© Springer-Verlag Berlin Heidelberg 2007
Printed in Germany

Typesetting: Camera-ready by author, data conversion by Scientific Publishing Services, Chennai, India
Printed on acid-free paper   SPIN: 12181247   06/3180   5 4 3 2 1 0

Preface

The Dagstuhl Seminar 05151 "Annotating, Extracting and Reasoning about Time and Events" took place April 10–15, 2005 at the International Conference and Research Center (IBFI), Schloss Dagstuhl, Germany. During the seminar, 17 leading researchers from 5 different countries presented current research and discussed open problems concerning annotation, temporal reasoning, and event identification. The work presented at this seminar, together with other previous and ongoing research, centers around an emerging *de facto* standard for time and event annotation: TimeML. TimeML has recently been adopted as a candidate for an ISO standard, and is currently being reviewed in this capacity.

At the seminar, the discussions focussed on the following three TimeML-related issues: using the TimeML language effectively for consistent annotation, determining how useful such annotation is for further processing, and describing modifications that should be applied to the standard for applications such as question-answering and information retrieval.

Discussions at the Dagstuhl Seminar led to new research ideas, and a variety of publications and conference and workshop presentations resulted. This current collection of papers adds to the growing body of work on TimeML. It focusses on important sub-areas within TimeML research such as temporal annotation and temporal reasoning and points to future research directions that are crucial for further progress.

The editors would like to thank participants for their attendance at the Dagstuhl Seminar, and for their contributions to the many lively and inspiring discussions. We are also grateful for having had Schloss Dagstuhl as a venue for this seminar. The research center is unique and provides a very nurturing environment for advancing exciting research. Finally, we would also like to thank Inderjeet Mani and David Ahn for providing additional reviews of several papers in this collection.

August 2007

Frank Schilder
Graham Katz
James Pustejovsky

Table of Contents

Annotating, Extracting and Reasoning About Time and Events

Frank Schilder[1], Graham Katz[2], and James Pustejovsky[3]

[1] R&D, Thomson Corp.
610 Opperman Drive, Eagan 55123, USA
Frank.Schilder@Thomson.com
[2] Institute for Cognitive Science University of Osnabrück
Kolpingstr. 7, 49076 Osnabrück, Germany
gkatz@uos.de
[3] Computer Science Department, Brandeis University
415 South St., Waltham, MA 02454 USA
jamesp@cs.brandeis.edu

Abstract. The main focus of the Dagstuhl seminar 05151 was on TimeML-based temporal annotation and reasoning. We were concerned with three main points: how effectively can one use the TimeML language for consistent annotation, determining how useful such annotation is for further processing, and determining what modifications should be applied to the standard to make it more useful for applications such as question-answering and information retrieval.

1 Introduction

Today's information extraction systems are capable of reliably extracting named entities such as PERSON or COMPANY names and LOCATIONS. Newspaper articles and other natural language texts, however, describe much more information between such entities than this. In particular, the underlying temporal relations between events would be very valuable for summarization system that produced summaries of developing stories. In order to provide a summary of a developing news story, for example, sequences of events need to be presented in a chronological and coherent way. A system that can produce such a summary would need to extract a time stamp for each event and to order the extracted events according to the time line. Current summarization systems are not able to do this or can offer only a rough approximation of the temporal information.

Such a summarization system is one example of future IE system that would require reliable temporal information in order to allow for temporal reasoning capabilities. Other systems that would benefit from temporal information include: Question-Answering systems, medical documentation systems, and legal reasoning systems.

Hence, a crucial first step toward the automatic extraction of information from texts is the capacity to identify what events are being described and to make explicit when these events occurred and which temporal relations hold

F. Schilder et al. (Eds.): Reasoning about Time and Events, LNAI 4795, pp. 1–6, 2007.

among them. There has recently been a renewed interest in making use of this kind of temporal and event-based information, with a wide variety of proposals and applications having been presented at recent conferences and workshops. [1,2,3,4,5].

Extracting temporal information from natural-language text is not trivial, since much of the temporal information conveyed in a natural language text is left implicit. Significant recent work has focused on developing schemata for making this information explicit, typically via annotation. An important result of contemporary research has been the adoption of a *de facto* standard for time and event annotation: TimeML [3,4,6].[1] This XML-based markup language is specifically designed for annotating texts with tags that make explicit the temporal and event-based information conveyed by the text, and has been adopted by a number of researchers in this domain. Much of our seminar was concerned with issues specific to this annotation scheme.

There are three basic types of tags used within the TimeML language:

TIMEX tags are used to annotate temporal expressions and provide them with a normalized value (e.g.,

```
<TIMEX tid="t1" val="2005-04-21">April 21st, 2005</TIMEX>
```

EVENT tags are used to annotate event expressions, providing "hooks" to relate them to other events and times:

```
<EVENT eid="ei">opened</EVENT>
```

TLINK tags indicate the temporal relations that hold between times and events (e.g. *the stock market opened on April 21st, 2005 at 10:00pm*):

```
<TLINK event="e1" relatedTime="t1" relation="IS_INCLUDED"/>
```

Other tags are used to capture more subtle semantic relations. SLINK tags, for example, are used to indicate various kinds of subordination relations, such as *reported speech*, in *The spokesman said the bomb injured 20 people*, or intensional contexts, such as in *Investors hoped that the stock market would open on April 21st, 2005 at 10:00pm*. Finally, ALINK tags indicate the aspect (or phase) of an event, as in *The market began to fall suddenly*. A corpus of 183 TimeML annotated documents (*TimeBank*) has been released by the LDC, and can be browsed and downloaded at timeml.org.

The seminar took place at Schloss Dagstuhl from 10. April - 15. April 2005. The central goal of the seminar was to consolidate the insights that have been made in recent years and to identify and address issues concerning annotation, temporal reasoning, and event identification that remain unresolved.

The various talks presented ranged from addressing the logical foundations of temporal reasoning to discussing the practical aspects of computing temporal information:

---

[1] To promote TimeML as a more formal standard, it has recently been adopted as a candidate for an ISO standard, and is currently being reviewed in this capacity.

1. Branimir Boguraev (IBM Research, USA)
   TimeBank-driven TimeML Analysis
2. Frank Schilder (Thomson R&D, USA)
   Temporal Information Extraction from Legal Documents
3. Andrea Setzer (University of Sheffield, GB)
   TimeML in a Medical Application
4. Jerry Hobbs (USC/ISI - Marina del Rey, USA)
   A Temporal Ontology for the Semantic Web
5. Lauri Karttunen and Annie Zaenen (PARC - Palo Alto, USA)
   Veridicity and Commitment?
6. Laure Vieu (LOA -Trento, I)
   Scope of Temporal Adverbials in Discourse
7. David Ahn (University of Amsterdam, NL)
   Towards Task-based Temporal Extraction and Recognition
8. Benjamin Han (CMU - Pittsburgh, USA)
   Understanding Times: An Constraint-based Approach
9. Tom Bittner (Univ. des Saarlandes, D)
   Approximate Qualitative Temporal Reasoning
10. Mark Steedman (University of Edinburgh, GB)
    The Calculus of Affordance
11. Marc Verhagen (Brandeis University, USA)
    Drawing TimeML Relations
12. Hans-Jürgen Ohlbach (Universität München, D)
    Computational Treatment of Temporal Notions the CTTN System
13. Rob Gaizauskas (University of Sheffield, GB)
    Getting Closure: Vagueness and Disjunction in TimeML
14. Graham Katz (University of Osnabrück, D)
    The Semantics of TimeML
15. James Pustejovsky (Brandeis University, USA)
    Event Arguments in TimeML
16. Ian Pratt-Hartmann (Manchester University, GB)
    Temporal Prepositions and their Logic
17. Inderjeet Mani (MITRE, USA)
    Chronoscopes: A theory of Underspecified Temporal Relations

This book contains selected papers that are further developments of the work presented at the workshop. The papers are representative for a set of important sub-areas identified by the seminar where progress has to be made in order to advance this field.

Issues Concerning Temporal Annotation. An increasingly important research question is concerned with the representation of temporal information, either while carrying out the annotation or for the purpose of representing it. Firstly, the annotation task can be made more reliable if the annotated temporal relations are easily viewable without burdening the annotator with too many details. Secondly, the resulting temporal information needs to be presented to a viewer in a understandable way. Previous work using a graph representation were already a step forward compared to simply adding temporal information to the XML tags. However, a graph annotation used for TANGO could produce too many links, because all implicit temporal information would be made explicit. In oder to deal with this representational problem, Marc Verhagen proposes a graphical representation called TBox that leaves some temporal information implicit by reducing the number of explicitly presented relations. This representation reduces the number of inconsistencies that may be introduced by a more cluttered representation of temporal information in an annotation tool.

Linguistic Analysis of Temporal Expressions. In this volume, Janet Hitzeman provides an account on the semantics of initial and sentence final temporal adverbials. She compares four different texts in order to evaluate whether text-type has an effect on the position (and interpretation) of temporal adverbials.

At the seminar, Laure Vieu presented an analysis of Locating Adverbials (LAs) such as *un peu plus tard* or *ce matin* (a little later, this morning) when they are dislocated to the left of the sentence (IP Adjuncts cases). She showed evidence that LAs seem to play an important part in structuring discourse. Lauri Karttunen and Annie Zaenen illustrate some cases of conventional implicature and show how they indicate an author's commitment to the truth of his/her statements and briefly state the importance of these distinctions for Information Extraction.

Learning from Annotations. Since the compilation of TimeBank, work on automatically learning the temporal relations has been enabled. Learning the temporal relations between the time stamp and events as well as between adjacent events in a text have recently been investigated within the SemEval competition. The TempEval task used the TimeBank corpus [5].

Within this collection, Bran Boguraev and Rie Kubota Ando present an in-depth analysis of TimeBank and discus experimental results on TimeML-compliant parsing via a blend of finite-state approaches with machine learning techniques.

Similarly, David Ahn, Joris van Rantwijk and Maarten de Rijke [7] published a follow-up paper of David Ahn's talk at the Dagstuhl seminar describing tagging temporal expressions via a cascaded approach combining several machine learning classifiers. Their experiments on the TERN 2004 data show that the cascaded machine-learning approach requires a much smaller number of composition rules for the derivation of the ISO-time stamps than competing approaches with comparable results.

New Domains. Talks at the seminar also identified interesting new domains for time and event annotation. Frank Schilder's paper discusses what kind of legal documents (legal narratives, transactional documents, statutes) may benefit from temporal information extraction and presents a prototype for extracting temporal information from U.S. statutes.

At the seminar, two other domains were discussed. Andrea Setzer presented a project from the medical domain.[2] In this project temporal information from patient notes dictated by doctors is to be extracted and mapped onto a database containing records of interventions (e.g., surgery) and investigations (e.g., X-RAY) performed on the patient. Ben Han investigated another domain that is different from the standard news texts TimeBank consists of (viz., email messages). Subsequent work by him was published at NAACL 2006 [8].

Time Logic. Practical tools for reasoning with temporal information were presented at the seminar by Benjamin Han and Hans-Jürgen Ohlbach. They presented implementations that do reasoning with temporal information, such as computing the current date "plus 2 months" (*two months from today*). Their implementations are written in Python (Han) and C++ (Ohlbach). Ohlbach expands on this work in the paper in this volume.

Graham Katz's paper provides a model-theoretic semantics for TimeML, closely based on Discourse Representation Theory. He addresses the problems of semantic scope, providing a second-order semantics that simulates semantic scope, and presents a very basic treatment of some of the non-extensional aspects of TimeML, namely the modality and the SLINK tags.

Ian Pratt-Hartmann published a paper based on his talk at the Seventh International Workshop on Computational Semantics [9]. His paper is concerned with the translation from TimeML annotation to temporal interval logic. Jerry Hobbs' presentation at the workshop was concerned with the annotation of durations of event descriptions in text. This work was published subsequently at different conferences and workshops (e.g., [10]). Mark Steedman analyzes temporal semantics for natural language in terms of a calculus developed for planning and reasoning about action. He proposed an event calculus based on Linear Dynamic Logic, and on instantaneous changes rather than intervals.

Reasoning. An important next step in this research area will involve techniques for reasoning with temporal information.

James Pustejovsky, Jessica Littman and Roser Saurí discuss the issue of whether TimeML should incorporate all of a verb's arguments into the markup specification language. They propose that the language of TimeML should make reference only to event arguments, and not to all verbal arguments. TimeML already makes reference to considerable argument structure in subordinating and aspectual contexts. These event-event relations between the predicate and an argument cover a large number of the events selected for by predicates. Most of those not covered, it is argued, are lexical discourse markers, such as *lead to*, and should be handled by a new LINK-type, called a DLINK (discourse link).

Inderjeet Mani's paper focuses on an important component of temporal reasoning that has been largely neglected: granularity. The author introduces an abstract device called chronoscopes that allows temporal abstraction over events and temporal relations depending on the chosen time granularity.

---

[2] http://nlp.shef.ac.uk/clef/

Another talk during the seminar was also concerned with granularity. Tom Bittner describes representation and reasoning methods taking the limits of our knowledge explicitly into account. For example, *happened yesterday* does not mean that x started at *12 am* and ended *0 pm*. He proposes an approach that describes the temporal location of events and processes as approximate and "rough" in nature, rather than exact and crisp. At the seminar, Rob Gaizauskas also discussed different approaches to representing the temporal information encoded in TimeML. He investigates how vague temporal information can be presented.

References

1. Mani, I., Wilson, G.: Robust temporal processing of news. In: Proceedings of the 38th Annual Meeting of the ACL, Hong Kong, Association for Computational Linguistics (2000)
2. Harper, L., Mani, I., Sundheim, B. (eds.): Proceedings on the Workshop on Temporal and Spatial Information Processing, Toulouse, France, ACL (July 2001)
3. Pustejovsky, J.: Terqas: Time and event recognition for question answering systems. In: ARDA Workshop, Boston, Mitre (2002)
4. Pustejovsky, J., Mani, I., Belanger, L., Boguraev, B., Knippen, B., Littman, J., Rumshisky, A., See, A., Symonen, S., Van Guilder, J., Van Guilder, L., Verhagen, M.: Arda summer workshop on graphical annotation toolkit for timeml. Technical report, Mitre, Boston (2004),
   `http://nrrc.mitre.org/NRRC/TangoFinalReport.pdf`
5. Verhagen, M., Gaizauskas, R., Schilder, F., Hepple, M., Katz, G., Pustejovsky, J.: Semeval-2007, task 15: Tempeval temporal relation identification. In: Proceedings of the Fourth International Workshop on Semantic Evaluations (SemEval-2007), Prague, Czech Republic, Association for Computational Linguistics, pp. 75–80 (June 2007)
6. Saurí, R., Littman, J., Gaizauskas, R., Setzer, A., Pustejovsky, J.: TimeML annotation guidelines version 1.2. Technical report, Brandeis University (January 2006),
   `http://www.ldc.upenn.edu/Catalog/CatalogEntry.jsp?catalogId=LDC20006T08`
7. Ahn, D., van Rantwijk, J., de Rijke, M.: A cascaded machine learning approach to interpreting temporal expressions. In: Human Language Technologies: The Conference of the North American Chapter of the Association for Computational Linguistics. Proceedings of the Main Conference, Rochester, New York, Association for Computational Linguistics pp. 420–427 (April 2007)
8. Han, B., Gates, D., Levin, L.: Understanding temporal expressions in emails. In: Proceedings of the Human Language Technology Conference of the NAACL, Main Conference, New York City, USA, Association for Computational Linguistics pp. 136–143 (June 2006)
9. Pratt-Hartmann, I.: From TimeML to Interval Temporal Logic. In: Proceedings of the 7th International Workshop on Computational Semantics, Tilburg, The Netherlands (January 2007)
10. Pan, F., Mulkar, R., Hobbs, J.R.: Learning event durations from event descriptions. In: Proceedings of the 44th Annual Meeting of the Association for Computational Linguistics (COLING-ACL), Sydney, Australia, Association for Computational Linguistics, 38–45 (2006)

Drawing TimeML Relations with TBox

Marc Verhagen

Computer Science Department
Brandeis University
Waltham, USA
marc@cs.brandeis.edu

Abstract. TBox is a new way of visualizing the temporal relations in
TimeML graphs. Until recently, TimeML's temporal relations were presented as rows in a table or as directed labeled edges in a graph. Neither mode of representation scales up nicely when bigger documents are
considered and both make it harder than necessary to get a quick picture of the temporal structure of a document. TBox uses left-to-right
arrows, box-inclusions and stacking as three distinct ways to visualize
precedence, inclusion and simultaneity.

Keywords: TimeML, temporal annotation, visualization, timelines,
temporal closure.

1 Introduction

In the early days of TimeML, the TimeBank corpus was created as an illus-
tration of the temporal annotation proposed by TimeML.[1] The first version of
TimeBank was annotated almost exclusively with the Alembic Workbench [6].
Alembic proved useful for annotation of non-relational tags, but it does not deal
neatly with the highly relational and inter-connected information embodied in
the temporal links (TLinks) of TimeML. In Alembic, TLinks can be added as
rows to a table where the columns denote the events and times that are linked
and the relation type of the TLink (before, after, includes etc). This works fine
when an annotator sweeps through the text linearly and creates TLinks between
events and times that are close to each other in the text. It makes it impossible,
however, to get a picture of what the temporal structure of a document is. In
addition, Alembic annotation of TLinks proved to be sensitive to directionality
errors where an annotator would, for example, accidentally add [X before Y]
instead of the intended [Y before X].

In 2003, a new tool named Tango [5,7] was developed in order to make
TimeML annotation more intuitive. Tango is a graphical annotation tool that
uses a graph to display the various links in a TimeML document. Annotation

---

[1] See [1,2] for an overview of TimeML and [3] for a description of TimeBank. TimeML
and TimeBank were created in the context of the ARDA/AQUAINT workshops
TERQAS and TANGO [4,5].

F. Schilder et al. (Eds.): Reasoning about Time and Events, LNAI 4795, pp. 7–28, 2007.

was expected to be more intuitive because with Tango it involves direct manipulation of a timeline. And indeed, adding a TLink does not involve elaborate
manipulation of a table, but proceeds by drawing arrows between events and
times that are displayed on a two-dimensional pane, as shown in figure 1. Temporal relations are represented by labels on the edges: precedence relations are
represented by arrows, proper and non-proper inclusion by circles, simultaneity
by boxes, and begins and ends by diamonds. Time expressions are printed in a
different color towards the top of the display, thereby suggesting a timeline. The
Tango graph only prints events and time expressions. Another Tango widget
contains the text where events and time expressions are color-coded. Selecting
an event in the Tango graph will highlight it in the text widget.

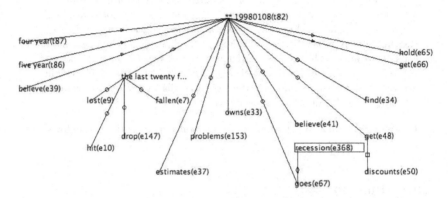

Fig. 1. A fragment from TimeBank Document ABC19980108.1830.0711, as displayed
by Tango. This is a black and white version of Tango's colored interface.

It turned out that Tango made annotation of TLinks more reliable and that
it invitedthe annotator to explore the temporal structure of a document more
thoroughly. Annotation with Tango also appears to result in a TimeML graph
where the events and times are more tightly connected.

There are a couple of limitations though. The main problem is that it is still
hard to quickly capture the temporal structure of the document. And when a
lot of links are involved the Tango display is simply not that clear. Larger documents can contain hundreds of links and graph clutter makes it hard to see the
big picture. There are several reasons for these difficulties. One is that all semantics pertaining to the kind of temporal relation is encoded on the label of the
graph and it is hard to make all the distinctions used by TimeML; using color
to group similar kinds of TLinks is not an option since color is already used
to distinguish between temporal links and other kinds of links like aspectual
links and subordination links. But most importantly, the visualization problems exist because no clear semantics is associated with the relative positions of
events and times with respect to each other. The annotator has complete free-
dom to place events and times where she likes them to be. Typically, some kind of

left-to-right ordering that follows precedence relations is adopted but one cannot rely on that. In many cases, the Tango display is only readily interpretable by the annotator who created it. Another drawback of the Tango display is that it does not discourage the annotator from adding TLinks that are inconsistent with existing TLinks. Given existing links [X before Y] and [Y before Z], only careful inspection of the labels will prevent an annotator from adding a link [Z before X].

This paper presents a better way to visualize TimeML relations. One that makes it easier to grasp the temporal structure of a document and one that, when embedded in an annotation tool, makes it less likely that inconsistencies are introduced. Section 2 introduces the TBox representation and its main display rules. Section 3 outlines the procedure that takes a TimeML annotation and visualizes it with TBox drawings. Section 4 dwells on the relation between annotation consistency and TBox visualization, sketching an informal proof that a consistent TimeML annotation can be drawn and that any TBox representation that can be drawn is a consistent annotation. Finally, section 5 explores the potential use of disjunctions in TimeML and its visualization.

2   Drawing TimeML Relations with TBox

A TBox is a graphical representation of events and times where temporal relations between the events and times are represented by left-to-right arrows, box inclusion and box stacking. The name TBox stands for temporal box or time box and it should be pointed out right here at the onset that there is no relation with the terminological box from Description Logic.

The central idea of TBox representations[2] is that relative placement of two events or times is completely determined by the temporal relations between them. Each event or time expression is placed in a box called a TBox. A TBox has a default size, but can be stretched as needed. Events and times have a similar ontological status in the sense that both participate in TLinks and both are placed in boxes in the TimeML graph. But times are distinguished from events by color-coding them. As mentioned before, TBox uses arrows, box inclusion and stacking instead of the labeled edges of Tango. TimeML relations are associated with visualization constraints that have to be met in the display. There are three constraints that determine placement of two events or times relative to each other:

1. Precedence Constraint. If event X is before event Y, then X's box will always be displayed to the left of Y's box and there is a sequence of arrows that leads from one box to the other. X and Y are not necessarily displayed at the same vertical position. A variation of this theme is when X is immediately before Y (X meets Y, or, in TimeML terminology, X ibefore Y). In that case, the arrow is replaced by a line ending in a solid dot.

---

[2] I will be using the terms TBox, TBox representation and TBox drawing interchangeably.

(1)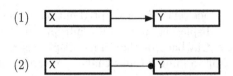

(2)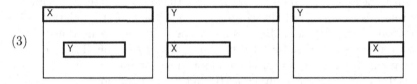

2. Inclusion Constraint. If X includes Y then the TBox of X is extended with a box that has thinner lines. The included event Y is placed inside this box. If needed, the including event X can be stretched so that it has space for Y. Y does not touch any side of the box. This constraint also governs begin and end relations. If Y begins or ends X, then the box of Y will touch the extension of X.

(3)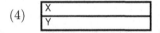

3. Equality Constraint. If X and Y are simultaneous then their boxes will be stacked directly on top of each other or there is a series of boxes between X and Y that are stacked similarly. If X and Y both include events, then these would be placed in a shared extension underneath Y. Simultaneity and identity are displayed differently. If two events are identical, then they will be placed together in the same box.

(4)

These constraints cover all TimeML relations. They assume an interval interpretation of TimeML events, similar to the one proposed by James Allen in [8]. If no rule governs placement of two events X and Y, then none of the configurations above will occur. X could be above, below, to the right or to the left of Y, but X cannot be inside the extension of Y, nor can there be a sequence of arrows between the two, nor can X and Y be stacked in any way.[3] It cannot be stressed enough that vertical and horizontal placement by themselves don't mean anything. They only mean something in connection with arrows, box inclusion or stacking.

Figure 2 shows the TBox representation of the TimeBank fragment that was displayed before in figure 1. At first blush, the TBox representation differs from the old Tango display in two important respects. There are fewer edges (6 as compared to 21), and temporal relations are expressed in more than one way. Not visible explicitly, but nevertheless important because it aides comprehension of the graph, is that this display is governed by the constraints given above. As

---

[3] Note that the constaints do not cover the overlap relation. TimeML currently has no overlap relation, but if it had, overlap could be represented as in section 5 where TBox representations of a larger set of temporal relations is given, including overlap in example (38).

a consequence, interpretating figure 2 is much easier than interpreting figure 1 and TBox representations are much more likely to reveal something odd in an annotation (figure 2, for example, may suggest that the annotator over-used the simultaneous relation a bit).

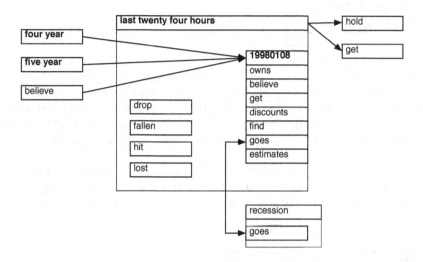

Fig. 2. A TimeBank fragment in TBox style

The TBox representation abandons the timeline metaphor. With a timeline, placement of an event e below a timeline element t strongly suggest that e is included in or overlapping with t. In reality, many events cannot be placed like this. The temporal structure of a document is not a timeline due to all kinds of under-specification that is prevalent in natural language. Having a timeline at the top of the display makes it hard to find a space for an event that is not related to any of the times in the timeline. The TBox representation has the added benefit that it is not attempting to visualize a timeline. Vertical placement under a time or date does not mean anything, unless inclusion or stacking is present. In figure 2, for example, there are no temporal relations between the four events grouped together inside of the large box. All we know is that they are included in the duration *last twenty four hours*.

2.1   Special Cases

The constraints above do need some elaborations. A special case occurs when one event is included in two unrelated events, that is, [X includes Z], [Y includes Z] and there is no TimeML relation between X and Y. The TBox way to represent this is to print Z twice and convey with an arrow that the two Z's are the same thing, as exemplified in (5).

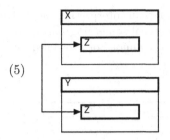

(5)

X and Y do not have to line up horizontally, but the two Z's do need to line up (and therefore X and Y will at least appear to have some overlap, which conforms to our intuitions). Any existing internal structure of Z will be displayed on only one of the two instances. There are related special cases for `begin` and `end` relations, as well as for certain mixes of `includes`, `begins` and `ends`.

The current formulation of the Precedence Constraint (placement of two events where one is before the other) fails to correctly account for the interplay of inclusion and precedence relations. Take the case where [X includes Y] and [X before Z].

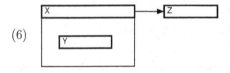

(6)

We display this case as above, but technically another arrow is needed from Y to Z because the display has strictly not made explicit that Y is before Z because there is no sequence of arrows between the two. However, the drawing above makes it completely obvious that [Y before Z]. So rule 1 should be amended, allowing that [X before Y] can also be expressed by X being included in a box that has a chain of arrows to Y.

The sister case is not problematic, but is worth looking at because it illustrates how certain inferences cannot and should not be made from a TBox drawing. If [X includes Y] and [Y before Z] then it cannot be inferred that [X before Z]. There should be no arrow from X to Z and the display should not strongly imply that X is before Z (it does strongly imply that X is at least not after Z).

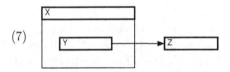

(7)

With this example, it becomes clear why the border of the extended box is thinner. If it were thicker, then it would be harder to imagine that Z is not necessarily after X. An alternative to the line would have been to use a grey area, but the problem with this approach is that it does not scale up neatly to a whole chain of inclusion relations.

The display rules by themselves do not force a minimal and clear representation of a TimeML graph. What they do is provide the basic building blocks for an intuitive visualization. The next section presents a procedure that creates a TBox drawing from a TimeML annotation.

3   A Procedure to Display TimeML Relations

There is a mechanical procedure to move from a TimeML annotation to a TBox drawing. It is not the case that every TimeML annotation can be mapped to a TBox drawing and the procedure will only successfully terminate for a subset of TimeML annotations, namely those annotations that are consistent. In fact, consistency checking is built into the procedure which consists of four stages: (1) temporal closure, (2) graph reduction, (3) mapping to an attribute-value matrix, and (4) mapping the attribute-value matrix onto a TBox drawing. These four steps are discussed in detail below.

3.1   Temporal Closure

The first stage is to create a TimeML annotation that is complete, where completeness means that any temporal relation that can be inferred from other relations is expressed by a TLink. This can be achieved using a constraint propagation algorithm as described in the interval algebra of James Allen [8]. Allen's algorithm also detects inconsistencies in the input, but unfortunately inconsistency checking in the unrestricted interval algebra[4] is not tractable. Tractability, however, is restored when a restricted version of interval algebra is used, based on the point algebra of Villain, Kautz and van Beek [9]. This restricted algebra is applied as a temporal closure component for TimeML in [10] and used as the first step in the display procedure. If an annotation is inconsistent then the procedure terminates at this stage. This happens often enough to warrant attention. In version 1.1 of the manually annotated TimeBank corpus, 32 out of 183 documents contain inconsistencies, and there is anecdotal evidence that an automatic temporal processing chain that does not include consistency checking produces even more inconsistencies.

3.2   Graph Reduction

The goal here is to map a complete TimeML graph to a unique minimal representation. Graph reduction consists of three sub steps: (1) creation of equivalence classes, (2) normalization of TimeML relations, and (3) deletion of relations that

---

[4] Unrestricted in the sense that any disjunction of relations is allowed as a label on the edges.

can be inferred. The first step reduces the number of nodes (and by extension the number of edges), the second step eliminates cycles, and the third step eliminates edges that can be derived from other edges.

1. Create Equivalence Classes

    The TimeML relations `identity`, `during`, and `simultaneous` are all equivalence relations. We can group events and times in equivalence classes and select one event or time to be the class representative. All TLinks from elements in the equivalence class to elements outside it are deleted except for relations from the class representative. Later, in the TBox drawing, this representative will be placed at the top of the box. The creation of equivalence classes is not without some ontological slight of hand. Equivalence relations are reflexive, symmetric and transitive and this is not the case for all three TimeML relations above. TimeML's `during` is not symmetric since an event is during a time and not the other way around. Also, TimeML does not stipulate anything about reflexivity and leaves it undefined. Using equivalence as a notion is valid however because we choose to interpret all TimeML relations as basic Allen relations between intervals. TimeML's `simultaneous`, `identity` and `during` are all mapped to `equals`, which is an equivalence relation.

2. Normalize Non-Equivalence Relations

    The procedure that turns an attribute-value matrix into a TBox drawing requires a directed acyclic graph with a limited set of temporal relations. Cycles can simply be removed by selecting a set of normalized relations and mapping the inverse relations to elements of the selected set. For example, [X after Y] can be mapped to [Y before X], [X is_included Y] to [Y includes X] and [X begins Y] to [Y begun_by X]. The set of normalized TimeML relations contains `before`, `ibefore`, `includes`, `begun_by`, and `ended_by`, as well as the equivalence relations `identity` and `simultaneous`.

3. Remove Implicit Relations

    The closure algorithm in [10] uses a complete set of compiled out composition rules. These rules can be used to delete relations that can be derived. For example, a `before` relation between X and Z can be removed given a composition rule like "`[X before Y]` ⊙ `[Y before Z]` = `[X before Z]`". The procedure uses two passes, one in which all derivable relations are marked and one that removes all marked relations. This reversed closure operation results in a unique minimal graph because all nodes that stand in equivalence relations to each other have been conflated into single nodes and therefore all potential non-deterministic cases have been eliminated. An example of a graph before and after this step is given in figure 3.

The reduced graph is semantically the same as the complete graph and allows the same inferences. The complete graph can be recreated from the minimal graph by reversing the steps in this section.

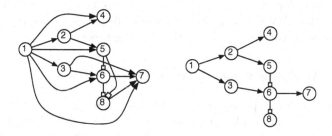

Fig. 3. The third step of graph reduction. Arrows indicate **before** links and lines that end in open squares indicate **includes** links.

3.3 From Graph to Attribute-Value Matrix

The minimal directed acyclic graph from the previous section can be trivially mapped to an attribute-value matrix (AVM) with re-entrancies and list values:

$$(8) \quad \begin{bmatrix} \text{id} & 1 \\ \text{before} & \left\langle \begin{bmatrix} \text{id} & 2 \\ \text{before} & \left\langle \begin{bmatrix} \text{id} & 4 \end{bmatrix}, \begin{bmatrix} \text{id} & 5 \\ \text{includes} & \boxed{6} \begin{bmatrix} \text{id} & 6 \\ \text{before} & \begin{bmatrix} \text{id} & 7 \end{bmatrix} \\ \text{includes} & \begin{bmatrix} \text{id} & 8 \end{bmatrix} \end{bmatrix} \right\rangle \end{bmatrix}, \begin{bmatrix} \text{id} & 3 \\ \text{before} & \boxed{6} \end{bmatrix} \right\rangle \end{bmatrix}$$

In general, every annotation can be represented in a single AVM, even those annotations that do not consist of one connected graph, but of a series of graphs. For those cases we need to introduce a root node that is not in the original annotation. This root node can be seen as the event before all other events and the roots of all other graphs will be pointed to from this root node with a **before** edge.

Let's print the AVM in example (8) slightly differently. The lists are flattened out by repeating the attribute name, node names are printed as an index, and boxes are drawn for clarity. The re-entrant box is shaded for clarity. These are simple mechanical changes but they make the following steps more transparent. The AVM above can now be printed as follows.

(9)
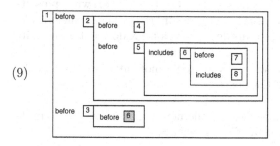

The main reason for undertaking this mapping from directed acyclic graph to AVM is that the AVM's visual layout is similar to a TBox drawing and that it facilitates an intuitive explanation of the drawing procedure.

3.4   From Attribute-Value Matrix to TBox

There is a bottom-up step-by-step process for replacing parts of a TimeML AVM with their corresponding TBox representations. Let's say we have an AVM, in which some parts have already been replaced by TBox drawings, as in the figure below.

(10)
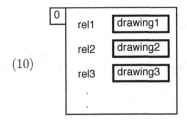

The process outlined below focuses on the **before** and **includes** relations, ignoring **ibefore**, **begun_by** and **ended_by**. These latter three relations act in ways very similar to **before** and **includes**.

Basic Mapping Rule. The basic AVM-to-drawing mapping rule for an AVM labeled 0 is as follows:

1. Draw a bar for 0 in the top left corner of the AVM
2. For every attribute equal to **before**, remove the attribute and draw an arrow from bar 0 to the head of the drawing that is the value of **before**. The head of the drawing is the bar that is at the top left, the one that dominates all others.
3. For every attribute equal to **includes**, remove the attribute and move the value of the attribute to the box underneath 0, draw the extension box if there isn't one yet. The value of the attribute should not be completely moved inside the box underneath, only the head of the drawing plus its extension should be moved in. This is to make sure that when [X includes Y] and [Y before Z], Z is not dragged inside of X because X does not necessarily include Z.
4. Remove the border of the box when there are no more attributes left in the box. Also remove the label of the box.

Application of the basic rule is illustrated in the next eight figures where parts of the AVM are gradually replaced by TBox drawings.

(11)

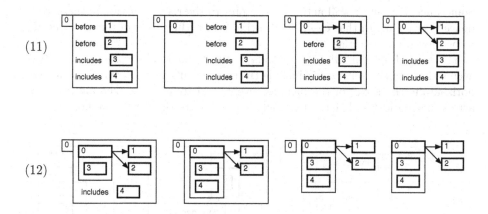

(12)

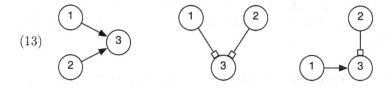

Special Cases. The picture is not complete yet because the basic rule only governs creation of drawings for those AVMs that do not include re-entrancies, that is, they work for trees, not for directed acyclic graphs. There are three special cases for nodes in the TimeML graph whose in-degree is higher than one:

(13)

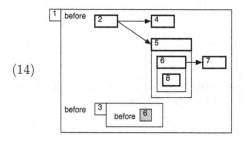

More generally formulated, the three special cases are (1) a node that has more than one incoming precedence relation from other nodes, (2) a node that has more than one incoming inclusion relation, and (3) a node that has a mixture of incoming precedence and inclusion relations. These three cases can be solved by using two extra rules: one special rule for merging in an arrow into a TBox that is already pointed to by another arrow or that is included in another box, and one special rule for merging two box inclusions. In most cases it is possible to use the first special rule and simply draw an arrow if one merges in a re-entrancy inside a **before** relation, as shown in example (14) below.

(14)

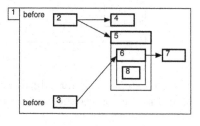

Sometimes, this arrow will actually not point to the right and violate the Precedence Constraint. In that case, the drawing that is the target of the arrow needs to be moved to the right.

The second special rule corresponds to the special rule in section 2.1: two events that are not related yet include the same event. The solution is to line up horizontally the boxes of the included event, and to add a connecting two-way arrow that indicates that the two included boxes are the very same box:

(15)

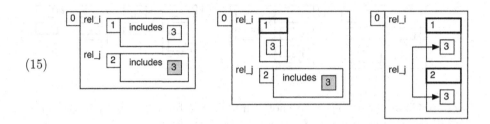

These two special rules require that the bottom-up redrawing strategy cannot indiscriminately take any attribute whose value is a TBox drawing and then apply the rules. Instead, care should be taken that the **includes** attributes are taken before the **before** attributes and that the re-entrancy is put on the **before** relation if possible.

A More Elaborate Example. The remainder of this section is devoted to a more extended example, showing how the AVM of section 3.3 is turned into a TBox drawing. The first two steps are to replace the sub-AVM labeled 6 with a box labeled 6, which encloses a box labeled 8 and connects with an arrow to a box labeled 7:

(16)

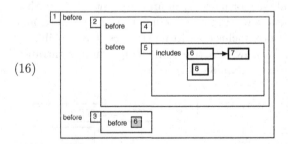

The next steps are to introduce another including box to represent that 5 includes 6, and to add precedence arrows from 2 to 4 and 5:

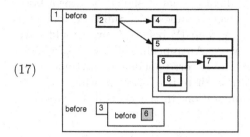

(17)

At this point, the only sub-AVM that can be drawn is the one with the re-entrant box labeled 6 in it. An arrow is now drawn from 3 to the target of the re-entrancy. Note that this requires that the target of the re-entrancy has already been drawn.

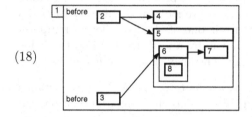

(18)

Finally, we take care of the last two remaining before links.

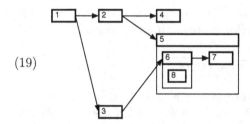

(19)

4   Consistency and Drawability

Temporal closure catches inconsistencies in an annotation. As does impossibility to arrange a graph with the rules above. Take the graph below, where each arrow indicates a before link.

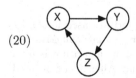

(20)

The Precedence Constraint of section 2 dictates that Z should be drawn both to the left of X and to the right of Y, which is impossible. Similarly, temporal closure will derive that given [X before Y] and [Y before Z], we should have [X before Z], but we already have [Z before X]. An appealing characteristic of the TBox representation is that it features a strong correspondence with annotation consistency.[5] For each consistent annotation there is a TBox drawing and each TBox drawing represents a consistent annotation. In this section, I will sketch an informal proof for these theorems.

Theorem 1. *Every consistent TimeML annotation can be visualized using a TBox representation.*

For simplicity's sake, I will sketch a related proof, namely that an attribute-value matrix that encodes a consistent annotation can be mapped to a TBox representation. That is, the steps from a consistent annotation to the AVM are taken for granted. The proof resembles an inductive proof and uses the procedure outlined in section 3.4. The bottom-up procedure looks at attribute-value pairs in an AVM labeled x where the attribute is one of the five allowed temporal relations (before, ibefore, includes, begun_by and ended_by) and the value is an empty AVM, a TBox drawing, or a re-entrancy. There are five cases.

1. The value is an empty AVM without attributes. This AVM, which stands for a single event or time expression with no outgoing temporal relation, can simply be drawn with a single TBox with label x and the attribute value pair can now be drawn as one of the TBox drawings in examples (1), (2) and (3) in section 2.

2. The attribute is before or ibefore and the value is a TBox drawing. This pair can simply be turned into a TBox representation by drawing an arrow from x to the embedded *TBOX*:

   (21) $\boxed{x}$ $\left[\text{before} \quad \boxed{TBOX}\right]$ $\implies$ $\boxed{x} \mapsto \boxed{\text{TBOX}}$

3. The attribute is one of includes, begun_by and ended_by and the value is a TBox drawing. This case is a bit more complicated because the internal structure of the TBox determines the action. If the head of the TBox has no precedence arrows leaving it, then the whole TBox is included in the new head. But if there is an outgoing before or ibefore relation, then only the head (plus anything in its extension box), will be included inside the extension of the new head.

   (22) $\boxed{x}$ $\left[\text{includes} \quad \boxed{x'} \mapsto \boxed{\text{TBOX}}\right]$ $\implies$

The previous cases together account for AVMs without re-entrancies, that is, for those annotation graphs that do not have nodes with more than one incoming edge. The most complicated cases deal with re-entrancies, that is, nodes that have more than one incoming edge (cf. example (13) in section 3.4).

4. The attribute of x is **before** or **ibefore** and the value is a re-entrancy named y. This case is isomorphic to the first special rule depicted in example (14) of section 3.4. It is important to note that in a consistent AVM the target of the re-entrancy y has to be outside of the AVM x because x and y can be linked by only one temporal relation. In addition, the target of y cannot be in $path_j$ because cycles were eliminated. This case can be schematized as follows:

(23) $\boxed{x^0}\begin{bmatrix} path_i & \boxed{TBOX} \\ path_j & \boxed{x}\begin{bmatrix} \text{before} & \boxed{y} \end{bmatrix} \end{bmatrix}$

Here, the value of $path_i$ is the TBox drawing that contains a TBox named y, and $path_i$ and $path_j$ are paths from the root of the AVM to the sub AVM named x and the TBox drawing. To create a drawing for the attribute-value pair inside of x, we draw an arrow from the box labeled x to the appropriate spot inside the TBox.

(24) $\boxed{x^0}\begin{bmatrix} path_i & \boxed{TBOX} \\ path_j & \boxed{x} \end{bmatrix}$

In this case, the arrow does not point to the left, which is a violation of the Precedence Constraint. This can be dealt with by shifting the TBox drawing to the right.

(25) $\boxed{x^0}\begin{bmatrix} path_i & \boxed{TBOX} \\ path_j & \boxed{x} \end{bmatrix}$

This shifting operation may set off a sequence of shifts because it is possible that from inside the TBox there is an arrow going to another part of the AVM named x^0. Note that this shifting process will always terminate because there are no cycles in the AVM.

To simplify the next case we are going to assume that re-entrancies are placed as the value of a **before** relation where possible. So the graph in example (26) will be represented as the AVM in (27) where the structure of the graph below node 3 is realized as the value of the **includes** relation and not the **before** relation.

(26)

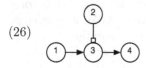

(27) $\boxed{x^0}$ $\begin{bmatrix} path_i & \boxed{1} \begin{bmatrix} before & \boxed{3} \end{bmatrix} \\ path_j & \boxed{2} \begin{bmatrix} includes & \boxed{3} \begin{bmatrix} before & \boxed{4} \end{bmatrix} \end{bmatrix} \end{bmatrix}$

With this assumption in place, we can continue with the last case.

5. The attribute of x is one of `includes`, `begun_by` and `ended_by`, and the value is a re-entrancy named y. This situation refers to the second special rule given in example (14) of section 3.4.

(28) $\boxed{x^0}$ $\begin{bmatrix} path_i & \boxed{TBOX} \\ path_j & \boxed{x} \begin{bmatrix} includes & \boxed{y} \end{bmatrix} \end{bmatrix}$

Given the assumption above, we know that the box labeled y inside of the TBox is either the head of the TBox, in which case $path_i$ ends in an `includes` relation, or is inside the TBox included in another box. In either case, this is an instance of example (5) in section 2.1 and the solution is to duplicate the box for y and draw a double arrow indicating that the two boxes refer to the same node. Example (29) shows the result for the case that the TBox consists of a box x' that includes box y and nothing else.

(29)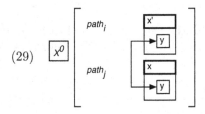

This concludes the proof of theorem 1. Note that we have again glossed over the `ibefore`, `begun_by` and `ended_by` relations. In section 3.4, when describing the mapping from AVMs to TBox drawings, we also ignored those relations. For the proof, their omission can again be motivated by the similarity of these relations to the once we did deal with. We now turn, in far less detail, to the reversed case.

Theorem 2. *Every TBox representation of a TimeML annotation is necessarily a representation of a consistent annotation.*

The proof is in many ways a reversed version of the proof of theorem 1. The idea is to use a bottom-up rebuilding of a TBox drawing into an AVM, and at each step show that the AVM has to be an AVM of a consistent annotation. We start with those parts of the TBox drawing where we have a TBox without extension and without outgoing arrows. These boxes can be replaced with empty AVMs and empty AVMs trivially are consistent. We then look at AVM's that are either included in one box or are pointed to from one box. The including box, or the box that points at the AVM can simply be incorporated, as shown for inclusion in example (30).

(30)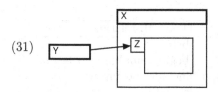

Note that if the embedded AVM named Y is consistent, then the new AVM has to be consistent as well. The more complicated cases are the ones where an AVM is not just included by one box or pointed to by one box, as in example (30).

(31)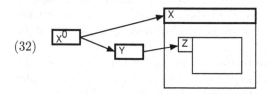

For a case like this, we need to create two AVM fragments, where one contains a re-entrance to the embedded AVM. This is best illustrated by embedding (30) in a larger AVM:

(32)

The corresponding AVM is as follows.

(33) $\boxed{x^0}$ $\begin{bmatrix} before & \boxed{X} \begin{bmatrix} includes & \boxed{Z} \begin{bmatrix} rel_i & val_i \\ rel_j & val_j \end{bmatrix} \end{bmatrix} \\ before & \boxed{Y} \begin{bmatrix} before & \boxed{Z} \end{bmatrix} \end{bmatrix}$

It is again the case that the new AVM is consistent, given the consistency of the AVM named Z. Nodes X and Y are not related to each other and having one include Z and the other be before Z does not introduce inconsistencies. Similar cases can be made for all other configurations.

5   Extending TimeML with Disjunctions

TimeML has defined a set of precise temporal relations along the lines of Allen's interval algebra, but some discussion has focused on whether more underspecified relations should be allowed. In the past, several authors (cf. [11], [12]) have reasoned that coarse temporal information is needed to properly describe indefinite temporal information in discourse, and within the TimeML community a discussion continues on what disjunctions of TimeML relations to allow.[6]

In this section, I consider a restricted set of disjunctions based on the point algebra proposed by Villain, Kautz and van Beek [9]. For this restricted set, we use precedence and equality relations on begin and end points of intervals. For example, we can define a relation between two intervals where the begin points are equal. This would translate into the following disjunction of TimeML relations: `simultaneous`, `begins` and `begun_by`. If we take $\{=, <, ?\}$ as the allowed temporal relations on points, then we can define a set of 29 convex relations[7], some examples are given below.

(34)

point constraints	Allen relations	TimeML relations
$x_2<y_1$	before	before
$x_1=y_1$	starts equals starts^i	begins simultaneous begun_by
$x_1<y_2 \wedge x_2>y_2$	during^i starts^i overlaps^i	includes begun_by
$x_1<y_1 \wedge x_2=y_2$	finishes^i	ended_by

This table shows constraints on point relations and the set of Allen relations that corresponds to the constraints. For example, the point relation $x_2<y_1$, where x_2 refers to the end point of the interval x and y_1 to the beginning point of interval y, corresponds to the `before` relation. Inverse relations are indicated with a superscript. TimeML translations of the Allen relations are given for comparison (there is no translation for overlap^i since overlap doesn't exist in TimeML).

With this expanded vocabulary of temporal relations, it is not the case anymore that drawability implies annotation consistency and inconsistent graphs can be drawn. As an example, take the annotation graph and TBox drawing in figure 4. The TBox drawing on the right is perfectly allowable given the constraints in section 2. But the annotation graph is not consistent. With closure we can compose [X {di,oi,si} Y] with [Y `fi` Z] and derive [X {di,oi,si} Z].

---

[6] It is not entirely the case that TimeML relations are all precise since some relations can be interpreted in a more liberal manner. For example, `simultaneous` and `during` can be seen as ambiguous between strictly `equals` in the Allen sense or as "mostly occurring in the same interval". For purposes of consistency checking and TBox representation these vague relations have been interpreted in a precise manner, which may in some cases not be conform to the annotator's interpretation. One could argue that there needs to be a distinction between precise and vague interpretations of these relations and that using disjunctions can be part of the solution.

[7] The term *convex* in this context means that the relations between intervals can be expressed as a conjunction of point relations.

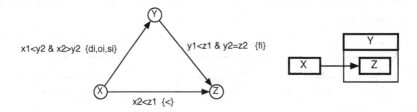

Fig. 4. An inconsistent annotation graph with its TBox drawing

Which is inconsistent with the already existing $[X < Z]$ because the intersection of {di,oi,si} with {<} is the empty set. The correspondence of annotation consistency and drawability is lost now the set of allowable temporal relations is expanded without adding visualization constraints. Of course, the only reason that there is a TBox drawing for the annotation graph is that there is no constraint that governs the disjunctive relation between X and Y. But the point is that with the current set of constraint we can create a TBox drawing for an inconsistent graph.

It turns out that the added disjunctive relations can all be easily displayed in a TBox-like fashion and that new visualization constraints can be formulated. Take for example the three disjunctions that describe how the begin points relate: $[x_1<y_1]$, $[x_1 = y_1]$, and $[x_1 > y_1]$. The first of these corresponds to the disjunction of TimeML relations [before $\vee$ ibefore $\vee$ ended_by $\vee$ includes]. The corresponding TBox drawings are depicted below.

(35)
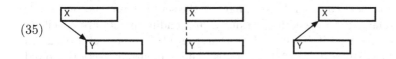

The main thing to note is that arrows are drawn from the corners of the boxes. Placement rules for arrows are unchanged, that is, the source of the arrow is placed to the left of the target of the arrow. A straight vertical dotted line connects two begin points that are equal. There are no restrictions on how the right corners of the boxes relate spatially, this is governed by the size of the boxes, which in turn is governed by the contents of the boxes. Three similar relations and TBox drawings can be defined for relations between end points. Other disjunctions occur when the beginning of one event precedes the end of the other and vice versa ($[x_1 < y_2]$, and $[y_1 < x_2]$), and when the beginnings of both events both precede the end of the other event ($[x_1<y_2 \wedge y_1<x_2]$).

(36)
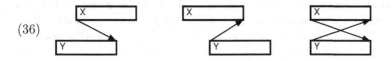

Six other disjunctions that are defined by two point relations are printed below without comment.

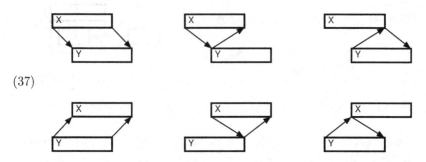

(37)

Those who were counting may have noticed that so far only 26 different disjunctions have been accounted for. The missing ones are (i) the totally underspecified relation, for which there exists no constraint, and (ii) the overlaps relation and its inverse, which does not exist in TimeML but which could be drawn as follows:

(38)

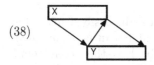

The set of 29 convex relations used above was generated by taking $\{=, <, ?\}$ as the allowed temporal relations between points. The set of 13 basic interval relations (and by extension, with some caveats, the set of TimeML relations), is a natural subset of these convex relations, namely the set of those relations where implicitly all relations between the beginning and ending points are specified. In the previous sections it became clear that the TBox representation is an attractive way to visualize TimeML relations. And the current section showed that every one of the 29 convex relations between events can be drawn and that the set of constraints in section 2 could be expanded accordingly.

Whether this should be done, and to what extent, is an empirical question depending on (i) simplicity of design, (ii) potential for increased clutter for each display relation, and (iii) added convenience and clarity of the display. For example, adding overlap to the display is unlikely to scale up gracefully when three or more events stand in overlap relations. On the other hand, adding lines between begin points may be a viable option.

5.1   Disjunctions and Consistency

Recall the example in figure 4 which showed that the constraints in section 2 allow you to draw inconsistent annotations. But the example given cannot be drawn if we expand our constraint set to include all 29 convex relations. The relation between X and Y now has a visual display, using arrows from the corners and an overlap of the horizontal extent:

(39)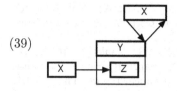

It is clear that X now has to be at two places at the same time and that the constraints that govern placement are now incompatible. So consistency and drawabilty are again flip sides of the same coin even with the expanded set of relations.

6 Conclusion

TBox is a viable and attractive alternative to the table-based and graph-based display modes of Alembic and Tango. A TBox representation is easier to read due to strict visualization constraints associated with temporal relations. A TBox representation is also always a representation of a consistent annotation. So if the annotation environment uses a TBox display then it is impossible to introduce inconsistencies. It should be stressed that I do not suggest that TBox representations should replace tables and graphs. A solid case can be made for an annotation environment where the annotator can switch freely between the modes, using the display mode that seems most comfortable at a given time. TBox has been implemented as an addition to Tango and will soon be made available at `http://www.timeml.org/tango/`.

Acknowlededgments

The TBox approach was inspired by an email from Nick Chubrich, who proposed many ways to improve on the Tango display. One of his ideas was to introduce a mechanism that allows annotators to select a whole group of events and use only one link to state that every event in this group stands in a particular temporal relation to another event or timex. TBox derives in a crooked way from this. Alex Baron implemented the TBox addition to the Tango annotation tool. Part of this work was carried out in the context of the AQUAINT TARSQI project and funded under US/DoD grant number NBCHC040027.

References

1. Pustejovsky, J.: Castaño, J., Ingria, R., Saurí, R., Gaizauskas, R., Setzer, A., Katz, G.: TimeML: Robust Specification of Event and Temporal Expressions in Text. In: IWCS-5 Fifth International Workshop on Computational Semantics (2003)
2. Pustejovsky, J., Knippen, R., Littman, J., Saurí, R.: Temporal and event information in natural language text. Language Resources and Evaluation 39, 123–164 (2005)

3. Day, D., Ferro, L., Gaizauskas, R., Hanks, P., Lazo, M., Pustejovsky, J., Saurí, R., See, A., Setzer, A., Sundheim, B.: The TimeBank Corpus. Corpus Linguistics (2003)
4. Pustejovsky, J., Belanger, L., Castaño, J., Gaizauskas, R., Hanks, P., Ingria, B., Katz, G., Radev, D., Rumshishky, A., Sanfilippo, A., Saurí, R., Setzer, A., Sundheim, B., Verhagen, M.: TERQAS Final Report. Technical report, The MITRE Corporation, Bedford, Massachusetts (2002)
5. Pustejovsky, J., Mani, I., Belanger, L., van Guilder, L., Knippen, R., See, A., Schwarz, J., Verhagen, M.: TANGO Final Report. Technical report, The MITRE Corporation, Bedford, Massachusetts (2003)
6. Day, D., Aberdeen, J., Hirschman, L., Kozierok, R., Robinson, P., Vilain, M.: Mixed-Initiative Development of Language Processing Systems. In: Fifth Conference on Applied Natural Language Processing Systems, Washington D.C., U.S.A., pp. 88–95 (1997)
7. Verhagen, M., Knippen, R.: TANGO: A Graphical Annotation Environment for Ordering Relations. In: Pustejovsky, J., Gaizauskas, R. (eds.) Time and Event Recognition in Natural Language. John Benjamin Publications (Forthcoming)
8. Allen, J.: Maintaining Knowledge about Temporal Intervals. Communications of the ACM 26(11), 832–843 (1983)
9. Vilain, M., Kautz, H., van Beek, P.: Constraint propagation algorithms: A revised report. In: Weld, D.S., de Kleer, J. (eds.) Qualitative Reasoning about Physical Systems, pp. 373–381. Morgan Kaufman, San Mateo, California (1990)
10. Verhagen, M.: Temporal Closure in an Annotation Environment. In: Pustejovsky, J., Gaizauskas, R. (eds.) Language Resources and Evaluation, vol. 39, pp. 123–164. Springer, Heidelberg (2005)
11. Freksa, C.: Temporal Reasoning Based on Semi-Intervals. Artificial Intelligence 54(1), 199–227 (1992)
12. Schilder, F.: Temporal Relations in English and German Narrative Discourse. PhD thesis, University of Edinburgh, Edinburgh, UK (1997)

Text Type and the Position of a Temporal Adverbial Within the Sentence*

Janet Hitzeman**

The MITRE Corporation
202 Burlington Road, M/S K309
Bedford, MA 01730 USA
hitz@mitre.org

Abstract. Consider example (a), below. When the temporal adverbial *since 1992* is in sentence-final position as in (i.a), it can attach syntactically at the VP-level or at sentence-level:

i. a. Mary has worked in Amsterdam since 1992.
   b. Since 1992 Mary has worked in Amsterdam.

Hitzeman (1993, 1997) argues that these different positions allow it to take on two readings: one in which there was some period between 1992 and speech time during which Mary worked in Amsterdam and another in which Mary has worked in Amsterdam for the period from 1992 until speech time. In contrast, sentence (i.b), in which the adverbial must attach at sentence-level, has only the second reading. If an initial-position adverbial unambiguously specifies the time of the event expressed by a sentence, then it should be a useful tool for a reader trying to determine the order of events in a narrative. To test the hypothesis that initial-position adverbials occur more often in texts describing events with some temporal order (i.e., a story line), I compare the use of these adverbials in narrative text and in non-narratives. The results show that significantly more initial-position adverbials are used in narratives. I then test the individual narratives and show that the significant difference in use of initial-position adverbials is correlated with the amount of flashback material in a narrative, i.e., with the complexity of the story line.

Keywords: temporal adverbial, narrative, perfect, flashback.

1 Introduction

In English, a phrase such as "a dog" can denote a specific entity in the world, e.g., Rover, or to any entity which fits the description "dog." Diesing (1992) shows that

---

* This work was done for ESPRIT Basic Research project DANDELION, funded by the European Union.
** The author's affiliation with The MITRE Corporation is provided for identification purposes only, and is not intended to convey or imply MITRE's concurrence with, or support for, the positions, opinions or viewpoints expressed by the author.

F. Schilder et al. (Eds.): Reasoning about Time and Events, LNAI 4795, pp. 29–40, 2007.

this phenomenon is common to many languages, and argues that it can be explained by positing that these phrases are interpreted differently at the syntax-semantics interface depending on their structural position in the sentence. Hitzeman (1993,1997) extends this argument to temporal adverbials with examples such as (1):

1. a. Mary has worked in Amsterdam for three years.
   b. For three years Mary has worked in Amsterdam.

When the temporal adverbial *for three years* is in sentence-final position as in (1a), it has two possible readings: one in which there was some three-year period in the past during which Mary worked in Amsterdam and another in which Mary has worked in Amsterdam for the three years preceding speech time. In contrast, sentence (1b) in which the adverbial is in sentence-initial position has only the reading in which Mary works in Amsterdam at speech time and has done so for the preceding three years.

   Obviously, there are other ways to precisely specify the time of an event, as in (2):

2. Mary worked in Amsterdam from November 5, 1995 until August 12th of the following year.

The question is whether authors make use of sentence-initial temporal adverbials, which pinpoint an event in time, as a tool to help the reader understand the order of the events described in the text. In order to explore this hypothesis, I examine the use of sentence-initial adverbials in four corpora from the European Corpus Initiative (ECI): *The Financial Times*, which contains descriptions of business dealings but not much of a story line, *Far from the Madding Crowd* which contains a simple forward-moving narrative, *A Christmas Carol*, which has one extended scene in which the main character visits his past, and *Silas Marner*, which has a great deal of flashback material throughout the text. The results will show that significantly more initial-position adverbials are used in narratives than non-narratives. A test of the individual narratives will show that the significant difference in use of initial-position adverbials is correlated with the amount of flashback material in a narrative, i.e., with the complexity of the story line.

2 The Ambiguity

The purpose of this section is to describe the two readings associated with a *for*-adverbial and to show that these temporal adverbials in English follow the pattern of having two readings when the adverbial is in sentence-final position but only one of these readings when it is in sentence-initial position.

   Consider the following examples:

3. a. Martha will be in her office for an hour.
   b. For an hour Martha will be in her office.

Sentence (3a) has two readings: The first reading is one in which Martha will be in her office for some unspecified hour in the future and another reading in which Martha will be in her office for the hour beginning at speech time. When the adverbial is in initial position, as in (3b), the only available reading is one in which Martha will be in her office for an hour beginning at speech time.

   I will adopt Klein's (1992) term position-definite (**p-definite**) to refer to an adverbial when its content fixes the position of a time span on the time axis, and I will call it **non-p-definite** when it is interpreted as expressing a time span whose position on the timeline is vague, e.g., we know that the NP refers to a one-hour interval after speech time, but we don't know the exact position of this interval. For example, *2:00 P.M. on January 9, 1983* is p-definite regardless of context since it refers to a particular time on the time axis, and expressions such as *yesterday* are p-definite in a context where the day of the utterance is known, while *two hours* can either be p-definite if it refers to a specific two hours or non-p-definite if it refers to any nonspecific or unknown two-hour period.

The contrast between the two readings in (3) is more clearly seen when the presence of another adverbial forces the non-p-definite reading, as below:

4. a. Martha will be in her office for an hour one day next week.
   b. Martha will be in her office one day next week for an hour.
   c. #For an hour Martha will be in her office one day next week.
   d. #One day next week for an hour Martha will be in her office.

The phrase *one day next week* forces the one-hour interval to be interpreted as being in the future so that the p-definite reading, in which the one-hour interval begins at speech time, is ruled out. The awkwardness of the sentence when the *for*-phrase is in initial position, as in (4c) and (d), shows that the non-p-definite reading is incompatible with the *for*-phrase in this position.

See Hitzeman (1997) for an extension of this argument to other English temporal adverbials.

3 The Effect of Text Type

3.1 The Initial Hypothesis

I have presented an analysis in which an adverbial in initial position must refer to a specific time, thus pinpointing the event described by the sentence in time, while an adverbial in final position is ambiguous between a p-definite and non-p-definite reading. How does this analysis of temporal adverbials play out in different text types? Consider that in a narrative the order of events is more important than in a fact-reporting, typically non-narrative text such as *The Financial Times (FT)*. In the bit of narrative in (5), for example, the events are understood to occur sequentially, and this is important for the understanding of the story:

5. *The Ghost conducted him through several streets familiar to his feet; and as they went along, Scrooge looked here and there to find himself, but nowhere was he to be seen. They entered poor Bob Cratchit's house; the dwelling he had visited before; and found the mother and the children seated round the fire. (Christmas Carol, line 6123)*

In the following texts from *The Financial Times* in (6), however, the order of events is less important:

6. a. *Thus organisations like British Rail and British Gas could use the wayleaves afforded by their railway lines and pipes to provide long-distance telecommunications. Cable TV operators would be allowed to provide local*

> *telecommunications services, which for 10 years BT would be excluded from doing. New competitors such as satellite operators would be allowed to operate international links. (FT, line 16490)*
>
> b. *The suggestions look sensible in themselves; but it does not seem likely that they will win credibility for more than a few months. The reason is political as much as economic: a free-enterprise government can hardly maintain a murderous squeeze on profits for ever, nor can a government nearing election choose to stick to rising unemployment. (FT, line 20462)*

More stative sentences and modals appear in these examples and there is no narrative progression. Aiding the reader in understanding how events progress in time is important to any cooperative author of a narrative. It is a reasonable hypothesis that the author of a narrative will use temporal expressions which are specific with respect to time, and that we will therefore find more temporal adverbials in initial position, where they are unambiguously p-definite, than in final position. I will test this hypothesis below.

3.2 The Raw Data

Before analyzing the data in these four texts, it was important that only temporal *for*-phrases which can be interpreted either as p-definite or non-p-definite should be included, since it is only those adverbials which can be disambiguated by moving them to sentence-initial position.

I first eliminated all non-temporal uses of *for*, such as *"For he could then see the path of his life clearly."* I also eliminated examples where the adverbial appeared in the middle of the sentence, as in (7), and examples in which the adverbial modified an NP, as in (8):

Table 1. Comparing *The Financial Times* with the three novels

		Financial Times		Three novels	
		initial	final	initial	final
for	p-def	5	36	21	32
	non-p-def	-	4	-	19
	Totals	5	40	21	51

Table 2. Comparison of the three novels

		A Christmas Carol		Silas Marner		Madding Crowd	
		initial	final	initial	final	initial	final
for	p-def	0	11	13	27	8	82
	non-p-def	-	2	-	1	-	16
	Totals	0	13	13	28	8	98

Table 3. χ^2 comparison of adverbial position for *The Financial Times* and the three novels

	Initial	Final	
The three novels	Observed: 21 E = (26 x 72) / 117 = 16 Contribution to χ^2 = 1.6	Observed: 51 E = (91 x 72) / 117 = 56 Contribution to χ^2 = 0.4	72
Financial Times	Observed: 5 E = (26 x 45) / 117 = 10 Contribution to χ^2 = 2.5	Observed: 40 E = (91 x 45) / 117 = 35 Contribution to χ^2 = 0.7	45
	26	91	117

χ^2 = 5.2
df = 1
Significant with p = .01

Table 4. χ^2 comparison of adverbial position for the three novels

	Initial	Final	
A Christmas Carol	Observed: 0 E = (21 x 13) /160 = 1.7 Contribution to χ^2 = 1.7	Observed: 13 E = (139 x 13) / 160 = 11.3 Contribution to χ^2 = 0.3	13
Silas Marner	Observed: 13 E = (21 x 41) / 160 = 5.4 Contribution to χ^2 = 10.7	Observed: 28 E = (139 x 41) / 160 = 35.6 Contribution to χ^2 = 1.6	41
Madding Crowd	Observed: 8 E = (21 x 106) / 160 = 13.9 Contribution to χ^2 = 2.5	Observed: 98 E = (139 x 106) / 160 = 92.1 Contribution to χ^2 = 0.4	106
	21	139	160

χ^2 = 17.2
df = 2
Significant with p = 0.01

7. *I have but to swallow this, and be <u>for the rest of my days</u> persecuted by a legion of goblins, all of my own creation. Humbug, I tell you; humbug!* (Christmas Carol, line 1315)
8. *Oil prices of Dollars 50 per barrel <u>for any length of time</u> could push the US and other economies into a fully-fledged recession. Some bond markets may have discounted a short, limited and successful armed conflict with Iraq, but a longer engagement would bring unforeseen pressures.* (FT, line 24854)

The analysis of the remaining *for*-phrases is shown in Tables 1 and 2. Table 1 shows the totals for The *Financial Times* and the three novels, and Table 2 shows the breakdown of the data for the three novels.

These phrases were analyzed as to whether they appeared in initial or final position, and also whether their interpretation in context was p-definite or non-p-definite. An example of such an adverbial is underlined in (9):

Table 5. Pairwise comparisons of the adverbials in the three novels

	Initial	Final	
A Christmas Carol	Observed: 0 E = (13 x 13) / 54 = 3.1 Contribution to χ^2 = 3.1	Observed: 13 E = (41 x 13) / 54 = 9.9 Contribution to χ^2 = 1.0	13
Silas Marner	Observed: 13 E = (13 x 41) / 54 = 9.9 Contribution to χ^2 = 1.0	Observed: 28 E = (41 x 41) / 54 = 31.1 Contribution to χ^2 = 0.3	41
	13	41	54

χ^2 = 5.4, df = 1
Significant with p = 0.05

	Initial	Final	
A Christmas Carol	Observed: 0 E = (8 x 13) / 119 = 0.9 Contribution to χ^2 = 0.9	Observed: 13 E = (111 x 13) / 119 = 12.1 Contribution to χ^2 = 0.1	13
Madding Crowd	Observed: 8 E = (8 x 106) / 119 = 7.1 Contribution to χ^2 =0.1	Observed: 98 E = (111 x 106) / 119 = 98.9 Contribution to χ^2 = 0.0	106
	8	111	119

χ^2 = 1.1, df = 1
Not significant

	Initial	Final	
Silas Marner	Observed: 13 E = (21 x 41) / 147 = 5.9 Contribution to χ^2 = 8.5	Observed: 28 E = (126 x 41) / 147 = 35.1 Contribution to χ^2 = 1.4	41
Madding Crowd	Observed: 8 E = (21 x 106) / 147 = 15.1 Contribution to χ^2 = 3.3	Observed: 98 E = (126 x 106) / 147 = 90.9 Contribution to χ^2 = 0.6	106
	21	126	147

χ^2 = 13.8, df = 1
Significant with p = 0.01

9. *It was one of his daily tasks to fetch his water from a well a couple of fields off, and for this purpose, ever since he came to Raveloe, he had had a brown earthenware pot, which he held as his most precious utensil among the very few conveniences he had granted himself. It had been his companion for twelve years, always standing on the same spot, always lending its handle to him in the early morning, so that its form had an expression for him of willing helpfulness, and the impress of its handle on his palm gave a satisfaction mingled with that of having the fresh clear water. (Silas Marner, line 822)*

Table 6. χ^2 shows significant results for *Silas Marner* and *The Financial Times*

	Initial	Final	
A Christmas Carol	Observed: 0 E = (5 x 13) / 58 = 1.1 Contribution to χ^2 = 1.1	Observed: 13 E = (53 x 13) / 58 = 11.9 Contribution to χ^2 = 0.1	13
Financial Times	Observed: 5 E = (5 x 45) / 58 = 3.9 Contribution to χ^2 = 0.3	Observed: 40 E = (53 x 45) / 58 = 41.1 Contribution to χ^2 = 0.0	45
	5	53	58

χ^2 = 1.5
df = 1
Not significant

	Initial	Final	
Silas Marner	Observed: 13 E = (18 x 41) / 86 = 8.6 Contribution to χ^2 = 2.3	Observed: 28 E = (68 x 41) / 86 = 32.4 Contribution to χ^2 = 0.6	41
Financial Times	Observed: 5 E = (18 x 45) / 86 = 9.4 Contribution to χ^2 = 2.1	Observed: 40 E = (68 x 45) / 86 = 35.6 Contribution to χ^2 = 0.5	45
	18	68	86

χ^2 = 5.5
df = 1
Significant with p = 0.05

	Initial	Final	
Madding Crowd	Observed: 8 E = (13 x 106) / 151 = 9.1 Contribution to χ^2 = 0.1	Observed: 98 E = (138 x 106) / 151 = 96.9 Contribution to χ^2 = 0.0	106
Financial Times	Observed: 5 E = (13 x 45) / 151 = 3.9 Contribution to χ^2 = 0.3	Observed: 40 E = (138 x 45) / 151 = 41.1 Contribution to χ^2 = 0.0	45
	13	138	151

χ^2 = 0.4
df = 1
Not significant

Out of context, the adverbial phrase *for twelve years* can be interpreted either as "for some twelve year period in the past" (the non-p-definite reading) or "for the twelve years preceding speech time" (the p-definite reading). In this context it takes on the latter reading.

There were also examples of *for*-phrases containing an NP the interpretation of which was necessarily either p-definite, as in (10), or non-p-definite, as in (11):

10. a. *"I'll be plain and open for the rest o' my life."* (*Silas Marner*, line 9057)

   b. *"Let it be, then, let it be," he said, receiving back the watch at last; "I must be leaving you now. And will you speak to me for these few weeks of my stay?"* (*Madding Crowd*, line 10643)

11. *The Soviet-backed government in Afghanistan declared an immediate amnesty for prisoners held   for up to three years to mark the fourth anniversary of a government drive for reconciliation.* (*FT*, line 858)

In (10) the phrases *the rest o' my life* and *these few weeks of my stay* refer unambiguously to particular time intervals in any context, and in (11) the phrase *for up to three years* must refer to an amount of time rather than a particular three year period. These types of examples are not of interest because their position in the sentence does not affect their interpretation. Similarly, I ignored examples with a generic interpretation such as (12):[1]

Table 7. χ^2 comparison of the number of *for*-phrases that occur in perfect sentences for the three novels

	Perfect	**No Perfect**	
A Christmas Carol	Observed: 7 E = (61 x 19) / 203 = 5.7 Contribution to χ^2 = 0.3	Observed: 12 E = (142 x 19) / 203 = 13.3 Contribution to χ^2 = 0.1	19
Silas Marner	Observed: 23 E = (61 x 54) / 203 = 16.2 Contribution to χ^2 = 2.9	Observed: 31 E = (142 x 54) / 203 = 37.8 Contribution to χ^2 = 1.2	54
Madding Crowd	Observed: 31 E = (61 x 130) / 203 = 39.1 Contribution to χ^2 = 1.7	Observed: 99 E = (142 x 130) / 203 = 90.9 Contribution to χ^2 = 0.7	130
	61	142	203

$\chi^2 = 6.9$, df = 2
Significant with p = 0.05

12. a. *It has been said that mere ease after torment is delight for a time; and the countenances of these poor creatures expressed it now. Forty-nine operations were successfully performed.* (*Madding Crowd*, line 8424)

   b. *So, when Priscilla was not with her, she usually sat with Mant's Bible before her, and after following the text with her eyes for a little while, she would gradually permit them to wander as her thoughts had already insisted on wandering.* (*Silas Marner*, line 8438)

In a generic sentence an adverbial can only have a non-p-definite interpretation, e.g., in (12b) *for a little while* doesn't refer to a single time but to a series of (non-p-definite) times during which Priscilla is not present. A present tense (non-reportative) sentence or a sentence with a frequency adverb such as *always* or a modal such as *can* are generic, and any adverbials in such sentences are likely to take on generic interpretations.

---

[1] See (Carlson, 1980) regarding generic sentences.

Table 8. χ^2 pairwise comparison of the number of perfect examples in the three novels

	Perfect	**No Perfect**	
A Christmas Carol	Observed: 7 E = (30 x 19) / 73 = 7.8 Contribution to χ^2 = 0.1	Observed: 12 E = (43 x 19) / 73 = 11.2 Contribution to χ^2 = 0.1	19
Silas Marner	Observed: 23 E = (30 x 54) / 73 = 22.2 Contribution to χ^2 = 0.0	Observed: 31 E = (43 x 54) / 73 = 31.8 Contribution to χ^2 = 0.0	54
	30	43	73

χ^2 = 0.2, df = 1
Not significant

	Perfect	**No Perfect**	
A Christmas Carol	Observed: 7 E = (38 x 19) / 149 = 4.8 Contribution to χ^2 = 1.0	Observed: 12 E = (111 x 19) / 149 = 14.2 Contribution to χ^2 = 0.3	19
Madding Crowd	Observed: 31 E = (38 x 130) / 149 = 33.2 Contribution to χ^2 = 0.1	Observed: 99 E = (111 x 130) / 149 = 96.8 Contribution to χ^2 = 0.1	130
	38	111	149

χ^2 = 1.5, df = 1
Not significant

	Perfect	**No Perfect**	
Silas Marner	Observed: 23 E = (54 x 54) / 184 = 15.8 Contribution to χ^2 = 3.3	Observed: 31 E = (130 x 54) / 184 = 38.2 Contribution to χ^2 = 1.4	54
Madding Crowd	Observed: 31 E = (54 x 130) / 184 = 38.2 Contribution to χ^2 = 1.4	Observed: 99 E = (130 x 130) / 184 = 91.8 Contribution to χ^2 = 0.6	130
	54	130	184

χ^2 = 6.7, df = 1
Significant with p = 0.05

3.3 The Results

The results showed a significant difference in use of initial-position adverbials between *The Financial Times* and the three narrative texts, as given in the χ^2 analysis in Table 3.[2] This result strengthens the argument that initial-position adverbials are used to help the reader interpret the temporal order of events in a narrative.

---

[2] χ^2 analysis was chosen because of its ability to accurately find significance for small amounts of data.

Table 9. χ^2 pairwise comparison perfect examples in *The Financial Times* with each of the three novels

	Perfect	No Perfect	
A Christmas Carol	Observed: 7 E = (27 x 19) / 106 = 4.8 Contribution to χ^2 = 1.0	Observed: 12 E = (79 x 19) / 106 = 14.2 Contribution to χ^2 = 0.3	19
Financial Times	Observed: 20 E = (27 x 87) / 106 = 22.2 Contribution to χ^2 = 0.2	Observed: 67 E = (79 x 87) / 106 = 64.8 Contribution to χ^2 = 0.1	87
	27	79	106

$\chi^2 = 1.6$
df = 2
Not significant

	Perfect	No Perfect	
Silas Marner	Observed: 23 E = (43 x 54) / 141 = 16.5 Contribution to χ^2 = 2.6	Observed: 31 E = (98 x 54) / 141 = 37.5 Contribution to χ^2 = 1.1	54
Financial Times	Observed: 20 E = (43 x 87) / 141 = 26.5 Contribution to χ^2 = 1.6	Observed: 67 E = (98 x 87) / 141 = 60.5 Contribution to χ^2 = 0.7	87
	43	98	141

$\chi^2 = 6.0$
df = 2
Significant with p = 0.05

	Perfect	No Perfect	
Madding Crowd	Observed: 31 E = (51 x 130) / 217 = 30.6 Contribution to χ^2 = 0.0	Observed: 99 E = (166 x 130) / 217 = 99.4 Contribution to χ^2 = 0.0	130
Financial Times	Observed: 20 E = (51 x 87) / 217 = 20.4 Contribution to χ^2 = 0.0	Observed: 67 E = (166 x 87) / 217 = 66.6 Contribution to χ^2 = 0.0	87
	51	166	217

$\chi^2 = 0$
df = 2
Not significant

Of course, even a financial text may have some small narrative elements. I will now argue that it is the complexity of the event ordering in a narrative that correlates with greater initial-position adverbial use, where a judgment of the amount of flashback material and a tally of the use of past perfects will serve as a measure of complexity.

First, note that the comparison of the narratives in Table 4 shows a significant difference in adverbial position in the three narratives. The pairwise comparison in

Table 5 then shows that *Silas Marner* stands out as significantly different from the others. A comparison of the individual novels with *The Financial Times*, given in Table 6, again shows that *Silas Marner* stands out as significantly different.

Consider that these narratives contain different amounts of flashback material, and that the interpretation of the order of the events described by such material is more complex than in a simpler, forward-moving narrative. The use of the perfect tense is a good indication that a sentence refers to an event in the past, such as the following passage from *Silas Marner*, where the fall is interpreted as occurring before speech time:

13. *Marner was highly thought of in that little hidden world, known to itself as the church assembling in Lantern Yard; he was believed to be a young man of exemplary life and ardent faith; and a peculiar interest had been centred in him ever since <u>he had</u> fallen, at a prayer-meeting, into a mysterious rigidity and suspension of consciousness, which, lasting for an hour or more, <u>had been</u> mistaken for death.*

A count of the number of perfects (past and present) in the texts can serve as an indication of how much flashback material they contain because they relate an event occurring at the *now* point with an event in the past, as in (14):

14. Silas picked up the shovel. He had buried the box.

One might pick up a shovel in order to bury something, but because of the perfect in the second sentence of (14) there is a reversal of the progression of events, and Silas must have picked up the shovel after burying the box.

The results in Table 7 show that there is a significant difference in the number of *for*-phrases in perfect sentences in the three novels. Breaking down the data to look at pairwise comparisons of the novels in Table 8 we see that *Silas Marner* has significantly more perfect examples than *Far from the Madding Crowd*. Similarly, as shown in Table 9, *Silas Marner* has significantly more perfect examples than *The Financial Times*, and *The Financial Times* doesn't show a significant difference in the number of perfect examples from either *A Christmas Carol* or *Far from the Madding Crowd*. More specifically, *Silas Marner* contains a considerable number of perfects (43% of the sentences with a *for*-phrase were in the perfect) as the main character, Silas, reflects on past phases of his life and goes back to the past to clear his name; *A Christmas Carol* has fewer (37%) since only one of the ghosts takes the main character, Scrooge, to the past; and *Far from the Madding Crowd* (24%) and *The Financial Times* (23%) contain even fewer since the former is a simple forward narrative and the latter is a non-narrative text which contains occasional passages containing narrative, but generally reports on factual, non-narrative matters.

4 Conclusion

Temporal adverbials such as *for an hour* can have two readings in sentence-final position but are unambiguous in initial position, fixing the event in time. The first hypothesis was that initial-position adverbials would be more common in narratives than in non-narratives. A comparison of three narratives and *The Financial Times* was consistent with this hypothesis. A second hypothesis was that more complex narratives would have significantly more initial-position temporal adverbials. The data was also consistent with this hypothesis; *Silas Marner*, a novel with many flashback scenes, stood out as having more sentence-initial adverbials than the other texts. A finer-grained categorization of "narrative" is therefore needed, and such a categorization may prove useful in temporal analysis of discourse.

References

1. Carlson, G.: Reference to Kinds in English. In: Hankamer, J. (ed.) Outstanding Dissertations in Linguistics, Harvard University Press (1980)
2. Diesing, M.: Bare Plural Subjects and the Derivation of Logical Representations. Linguistic Inquiry 3, 353–380 (1992)
3. Hitzeman, J.M.: Temporal Adverbials and the Syntax-Semantics Interface. Ph.D. thesis, University of Rochester, Department of Linguistics, Rochester, NY (1993)
4. Hitzeman, J.: Semantic Partition and the Ambiguity of Sentences Containing Temporal Adverbials. Journal of Natural Language Semantics (1997)
5. Klein, W.: The Present Perfect Puzzle. Language 68(3), 525–552 (1992)

Effective Use of TimeBank for TimeML Analysis

Branimir Boguraev and Rie Kubota Ando

IBM T.J. Watson Research Center, 19 Skyline Drive, Hawthorne, NY 10532, USA
bran@us.ibm.com, rie1@us.ibm.com

Abstract. TimeML is an expressive language for temporal information, but its rich representational properties raise the bar for traditional information extraction methods when applied to the task of text-to-TimeML analysis. We analyse the extent to which TimeBank, the reference corpus for TimeML, supports development of TimeML-compliant analytics. The first release of the corpus exhibits challenging characteristics: small size and some noise. Nonetheless, a particular design of a time annotator trained on TimeBank is able to exploit the data in an implementation which deploys a hybrid analytical strategy of mixing aggressive finite-state processing over linguistic annotations with a state-of-the-art machine learning technique capable of leveraging large amounts of unannotated data. We present our design, in light of encouraging performance results; we also interpret these results in relation to a close analysis of TimeBank's annotation 'profile'. We conclude that even the first release of the corpus is invaluable; we further argue for more infrastructure work needed to create a larger and more robust reference corpus.[1]

Keywords: corpus analysis, TimeBank, TimeML, temporal information extraction, machine learning.

1   Introduction

TimeML was designed [1] to connect the processes of temporal analysis of a text document into a rich, intermediate, representation and its subsequent formalisation by means of an ontology of time [2]. This paper assumes some familiary with TimeML; in essence, the language uses the representational principles of XML markup to annotate the analysis of the core elements in a temporal framework: time expressions, events, and links among these (additionally moderated by temporal connectives, or signals). For details of the markup language for time, readers are referred to [3].

In line with the established methodology of creating community-wide annotated resources, where linguistic analysis is captured by means of a range of tags, and finer-grained specification of analytical detail is expressed by means of suitably defined attributes on these tags, TimeML implements a flexible representational scheme for text markup. At the same time, the language takes

---

[1] This work was supported in part by the ARDA NIMD (Novel Intelligence and Massive Data) program PNWD-SW-6059.

F. Schilder et al. (Eds.): Reasoning about Time and Events, LNAI 4795, pp. 41–58, 2007.

the notion of markup to an extreme, developing half-a-dozen entity and relation marking tags—both consuming and non-consuming—and defining a large number of attributes for most of them.

Consequently, the resulting language is both very expressive and very complex. The expressiveness is almost a necessity, arising from the richness of time information and depth of temporal analysis, and addressed from the beginning of the design effort. The complexity is at least in comparison with markup schemes designed for the kinds of "named entities" which have traditionally been at the focus of conventional information extraction (IE) endeavours.

Many markup schemes for IE to date target relatively simple phenomena; unlike TimeML, their design has not been informed by the need to capture the variety and complexity of information required to support inference and reasoning. The extent to which IE can be argued to offer some basis for language understanding can be found in the 'spirit' of the MUC[2] *event scenario* tasks, which instantiate semantic networks [4,5]. However, the mapping of an entire text document to a single template can hardly be regarded as logically complete and coherent, in the sense required and assumed by formal event and/or time ontologies.

More recently, a growing body of work has initiated investigations into the nature of *linguistic annotation*—as a principled description of a linguistic phenomenon of interest (see, for instance, [6]). Such a description, of course, would then be instrumental to a deeper level of analysis and understanding.

Interestingly enough, an early instance of such an annotation effort—with a schema focusing on an identifiable linguistic phenomenon, and not just "named entities" markup—was defined by the ACE[3] event timestamping task, which sought to identify within-sentence event-time links. it could be argued that even within the limited set of event classes defined to be in the scope of ACE, the emphasis in schema design was on the annotation of relational information over a full inventory of temporal relations, and not just that of extents and spans.

In a similar spirit, TimeML aims to capture a much richer set of the temporal characteristics in a text document, so that the intricate temporal linking among all time expressions and events can then get fully mapped onto an ontologically-grounded temporal graph (or its equivalent), cf. [7], [8]. Indeed, such a mapping (see [9] for a sketch) has been one of the guiding principles in the conception and design of TimeML.

The design of TimeML therefore brings both promises—but also challenges—as its representational properties significantly raise the bar for traditional information extraction methods. A particularly relevant question, then, concerns the extent to which TimeML-compliant analysis can be automated: temporal reasoning frameworks crucially require such analysis for any practical understanding of time: "... the [TimeML] annotation scheme itself, due to its closer tie to

---

[2] Message   Understanding   Conferences;   see   `http://www.itl.nist.gov/iad/894.02/related_projects/muc/main.html`

[3] Automatic Content Extraction; see `http://www.nist.gov/speech/tests/ace/index.htm`

surface texts, can be used as the first pass in the syntax-semantics interface of a temporal resolution framework such as ours. The more complex representation, suitable for more sophisticated reasoning, can then be obtained by translating from the annotations." [8].

Analysis into TimeML is the primary question addressed by this paper. We start from the position stated by Pustejovsky et al. [10] as one of the guiding motivations for developing the TimeBank corpus, which is the primary reference resource for TimeML: it would be regarded as a resource for "training and evaluating algorithms which determine event ordering and time-stamping" (ibid.), as well as providing general-purpose training data for any and all TimeML components. We then demonstrate that small (and somewhat noisy) as it is (compared to guidelines implicitly established by other information extraction tasks relying on annotated data), TimeBank is still the valuable resource that [10] describes.

Our method rests on developing a strategy for time analysis of text specifically informed by the characteristics of TimeBank: a synergistic approach deploying both finite-state (FS) grammars with broad range of analysis and machine learning techniques capable of also leveraging unannotated data. Thus we aim to make maximal use of the information captured by this particular corpus, even if it was not explicitly designed and constructed as a proper training resource.

2 Quantitative and Qualitative Analysis of TimeBank

One of the common characteristics of annotation efforts is that they make, from the outset, infrastructural provisions for the development of a substantial 'reference' corpus, which defines a gold standard ("truth") for the task. The corpus contains materials selected to be representative of the phenomenon of interest; sizes of training and testing samples are carefully considered especially as they depend on the complexity of the task; experienced annotators are used; the corpus is not released until a certain level of inter-annotator agreement is reached. These measures ensure that the reference corpus is of a certain size and quality.

The TimeBank corpus is small. This need not be surprising, given that the TERQAS[4] effort did not commit to producing a 'reference', training-strength, corpus in the sense described above. In fact, TimeBank is almost a 'side effect' of the work: it was largely an exercise in applying the annotation guidelines— as they were being developed—to real texts (news articles, primarily) in order to assess the need for, and then the adequacy of, the language representational devices as they were being designed in the process of TimeML evolution.

Just how small TimeBank is is illustrated by the following statistics. The corpus has only 186 documents, with a total of 68.5K words. As there are no

---

[4] Temporal and Event Recognition for QA Systems; http: //www.timeml.org/ terqas/index.html). The TERQAS effort coordinated, over an extended period of time, a series of definitional and follow-up workshops from which emerged the current set of TimeML annotation guidelines.

separate training and test portions, it would need partitioning somehow; if we held out 10% of the corpus as test data, we have barely over 60K words for training.

To put this into perspective, this is order of magnitude less than other standard training corpora in the NLP community: the Penn Treebank corpus[5] for part-of-speech tagging (arguably a simpler task than TimeML component analysis) contains more than 1M words—which makes it over 16 times larger than TimeBank; the CoNLL'03 named entity chunking task[6] is defined by means of a training set with over 200K words. A task closely related to time analysis is ACE's TERN (Temporal Expression Recognition and Normalisation)[7]. TERN only focuses on TIMEX2 (TIMEX3, which extends the TIMEX2 tag [3], is just one of half-a-dozen TimeML components); even so, the TERN training set is almost 800 documents/300K words-strong.

Boguraev et al. [11] offer a detailed quantitative and qualitative analysis of the TimeBank corpus, in its original version—which was the basis for the experiments and results reported in this paper. In general, the observation is that the combination of the small size of TimeBank, the uneven distribution of TimeML components, and the erroneous annotation introduced by mixture of infrastructure issues and annotation methodology, lead to some significant challenges in using the corpus as a training resource.

Consider, for instance, the extreme paucity of positive examples over a range of categories. Fig. 1 (reproduced here, for convenience of reference, from [11]) shows the distribution of TLINK and EVENT types. These are the 'targets' of relational time analysis, capturing the temporal semantics above time expressions. As such, they are crucial for any analytical device.

The numbers in the figure illustrate the highly uneven distribution of this category data. The numbers also reveal some of the variety and complexity of TimeML annotation: the extensive typing of EVENTs, TIMEX3's and LINKs introduces even more classes in an operational TimeML typology. Thus an event recognition and typing task is, in effect, concerned with partitioning recognised events into 7 categories (as we shall see in Section 5.2, a particular implementation of such a partitioning is realised as $(2k + 1)$-way classification task, where $k = 7$ in our case). Similarly, for TLINK analysis the relevant comparison is to consider that in contrast to, for instance, the CoNLL'03 named entity recognition task—with training data containing 23K examples of named entities belonging to just 4 categories, TimeBank offers less than 2K examples of TLINKs, which, however, range over 13 category types.

The analysis in [11] additionally discusses the sources of noise in the first release of TimeBank. Broadly speaking, there are three different categories of error: errors due to failures in the annotation infrastructure, errors resulting from broad interpretation of the annotation guidelines, and errors due to the inherent complexity of the annotation task (further compounded by underspecification in

---

[5] See http://www.cis.upenn.edu/~treebank/home.html
[6] See http://cnts.uia.ac.be/conll2003/ner/
[7] See http://timex2.mitre.org/tern.html

TLINK type	# occurrences	EVENT type	# occurrences
IS_INCLUDED	866	OCCURRENCE	4,452
DURING	146	STATE	1,181
ENDS	102	REPORTING	1,010
SIMULTANEOUS	69	I_ACTION	668
ENDED_BY	52	I_STATE	586
AFTER	41	ASPECTUAL	295
BEGINS	37	PERCEPTION	51
BEFORE	35		
INCLUDES	29		
BEGUN_BY	27		
IAFTER	5		
IDENTITY	5		
IBEFORE	1		
Total :	1,451	Total :	8,243

Fig. 1. Distribution of (some) TimeML component types. Note that the count of 1451 TLINKs refers only to the TLINKs between an event and a temporal expression, itself in the body of a document. (TLINKs with TIMEX3's in metadata are not counted here.)

the guidelines). The reader is referred to that discussion, because it is important, for correctly situating our experiments and interpretating the results, to have an appreciation of the degree of noise which is at a level above what typically might be expected of a training resource.

Parenthetically, we observe that the kind of detailed analysis presented in [11]—itself motivated by the desire to understand how to interpret the performance figures reported in this paper—was itself the basis for a focused effort to revise and clean up the TimeBank corpus, which is currently distributed (as Version 1.2) through the offices of the Linguistic Data Consortium.

3   Challenges for TimeML Analysis

It is clear that temporal annotation is a very complex problem: TimeML was developed precisely to address the issues of complexity and to provide a representational framework capable of capturing the richness of analysis required. One consequence of this is the pervasiveness of relational data which is integral to the underlying representation: all links are, notationally, relations connecting events with other events or temporal expressions. As recent work in relation finding information extraction shows (in particular, in the context of the ACE program), the task requires both some linguistic analysis of text and the definition of complex learning models, typically going beyond just token sequences.

Additionally, as the previous section shows, a different degree of complexity is introduced by the size (and coverage) characteristics of TimeBank. While it may be reasonable to take a position that in our investigation we will focus on those TimeML components which are relatively more prevalent in the data (*e.g.* TLINKs over ALINKs and SLINKs), we still need to address the problem of insufficient training data. Our position thus is that in addition to deploying sophisticated feature generators, we crucially need to leverage machine learning technology capable of exploiting unlabeled data.

Our strategy for TimeML analysis of text develops a hybrid approach utilising both finite-state (FS) grammars over linguistic annotations and machine learning (ML) techniques incorporating a novel learning strategy from large volumes of unlabeled data. The respective strengths of these technologies are well suited for the challenges of the task: complexity of analysis, need for some syntactic and discourse processing, and relative paucity of examples of TimeML-style annotation.

The initial targets of our analysis are TIMEX3 (with attributes), EVENT (plus type), and TLINK (plus type, and limited to links between events and time expressions); see Section 2 and Fig. 1. This kind of limitation is imposed largely by the distributional properties of TimeML components annotated in TimeBank (as discussed in Section 2 earlier); but it is also motivated by the observation that to be practically useful to a reasoner, a time analysis framework would need to support, minimally, time stamping and temporal ordering of events. As this is work in progress, the description below offers more details specifically on identifying TIMEX3 expressions, marking and typing EVENTs, and associating (some of these) with TIMEX3 tags (typing the links, as appropriate).

All of these subtasks have components which can be naturally aligned with one or the other of our strategic toolkits. Thus TIMEX3 expressions are intrinsically amenable to FS description, and a grammar-based approach is well-suited to interfacing to the task of TIMEX3 normalisation (*i.e.* instantiating its value). On the other hand, certain attributes of a TIMEX3 (such as temporalFunction, valueFromFunction, functionInDocument) can be assigned by a machine learning component. FS devices can also encode some larger context for time analysis (temporal connectives for marking putative events, clause boundaries for scoping possible event-time pairs, *etc*; see Section 4). To complement such analysis, an ML approach can, using suitable classification methods, cast the problem of marking (and typing) EVENTs as chunking (Section 5.2). As we will see later, a TLINK classifier crucially relies on features derived from the configurational characteristics of a syntactic parse; a result in line with recent work which shows that mid-to-high-level syntactic parsing—typically derived by FS cascades—can produce rich features for classifiers.

In summary, we address the challenges of the TimeBank corpus by combining FS grammars for temporal expressions, embedded in a shallow parser adapted for time analysis, with machine learning trained with TimeBank and unannotated corpora.

4   Finite State Devices for Temporal Analysis

Temporal expressions conform to a set of regular patterns, amenable to grammar-based description. Viewing TIMEX3 analysis as an information extraction task, a cascade of finite-state grammars with broad coverage (compiled down to a single TIMEX3 automaton with 500 states and over 16000 transitions) targets abstract temporal entities such as unit, point, period, relation, *etc*; typically, these will be further decomposed and typed into *e.g.* month, day, year (for a unit); or interval or duration (for a period).

Temporal expressions are characterised by "local" properties—granularity (*e.g.* month, day, *etc*), cardinality, ref_direction(*e.g.* prior, or subsequent to "now"), and so forth—which are intrinsic to their temporal nature, but not directly related to TIMEX3 attributes. Fine-grained analysis of temporal expressions, instantiating such local propertiesis is crucially required for normalising a TIMEX3: consider, for example, that representing *e.g. "the last five years"* as illustrated in Fig. 2 below greatly facilitates the derivation of a value (in this case "5PY") for the TIMEX3 value attribute.

```
[timex : [relative       : true ]
         [ref_direction  : past ]
         [cardinality     : 5 ]
         [granularity     : year ] ]
```

Fig. 2. Analysis of a time expression in terms of local attributes

In effect, such analysis amounts to a parse tree under the TIMEX3. (Not shown above is additional information, anchoring the expression into the larger discourse and informing other normalisation processes which emit the full complement of TIMEX3 attributes—type, temporalFunction, anchorTimeID, *etc*).

It is important to separate the processes of recognition of the span of a TIMEX3 expression from local attribute instantiation for that expression. There is nothing intrinsic to the recognition which necessitates a grammar-based description in preference to a statistical model (as the TERN evaluation exercise demonstrated [12]). However, local attributes (as exemplified above) are necessary for the interpretation rules deriving TIMEX3 value.

TimeBank does not contain such fine-grained mark-up: the grammars thus perform an additional 'discovery' task, for which no training data currently exists, but which is essential for discourse-level post-processing, handling *e.g.* ambiguous and/or underspecified time expressions or the relationship between document-internal and document-external temporal properties (such as 'document creation time').

In addition to *parsing of temporal expressions*, FS devices are deployed for *shallow parsing for feature generation*. We build upon prior work [13], which showed how substantial discourse processing can be carried out from a shallow syntactic base, and derived by means of FS cascading.

Our grammars interleave syntactic analysis with named entity extraction. In particular, they define temporal expressions—as well as other TimeML components, namely events and signals—in terms of *linguistic* units, as opposed to simply lexical cues (as many temporal taggers to date do). The focus on linguistic description cannot be over-emphasised. One of the complex problems for TimeML analysis is that of event identification. A temporal tagger, if narrowly focused on time expressions only (cf. [14]), offers no clues to what events are there in the text. In contrast, a temporal parser aware of the syntax of a time phrase like *"during the long and ultimately unsuccessful war in Afghanistan"* is very close to knowing—from the configurational properties of a prepositional phrase—that the nominal argument (*"war"*) of the temporal preposition (*"during"*) is (most likely) an event nominal.

This kind of information is easily captured within a parsing framework. Additionally, given that EVENTs and LINKs are ultimately posted by a machine learning component, the parser need not commit to *e.g.* event identification and typing. It can gather clues, and formulate hypotheses; and it can then make these available to an appropriate classifier, from whose point of view an EVENT annotation is just another feature. Indeed, the only use of syntactic analysis beyond the TIMEX3 parser is to populate a feature space for the classifiers tasked with finding EVENTs and LINKs (Section 5).

Feature generation typically relies on a mix of lexical properties and some configurational syntactic information (depending on the complexity of the task). The scheme we use (Section 5) requires additionally some semantic typing, knowledge of boundaries of longer syntactic units (typically a variety of clauses), and some grammatical function. Fig. 3 illustrates the nature of the FS cascade output.

Most of the above is self-explanatory, but we emphasise a few key points. The analysis captures the mix of syntactic chunks, semantic categories, and TimeML components used for feature generation (a label like GrmEventOccurrence denotes a hypothesis, generated by the syntactic grammars, that *"earned"* is an occurrence type EVENT). It maintains local TIMEX3 analysis; the time expression is inside of a larger clause boundary, with internal grammatical function identification for some of the event predicates. The specifics of mapping configurational information into feature vectors is described in Section 5.

```
[Snt [svoClause
  [tAdjunct In [NP [timex3 the 1988 period timex3] NP] tAdjunct],
  [SUB [NP the company NP] SUB]
  [VG [GrmEventOccurrence earned grmEventOccurrence] VG]
  [OBJ [NP [Money $20.6 million Money] NP] OBJ] svoClause] ... Snt]
```

Fig. 3. Shallow syntactic analysis (simplified) from finite-state parsing

TimeML parsing is thus a bifurcated process of TimeML components recognition: TIMEX3's are marked by FS grammars; SIGNALs, EVENTs and LINKs are putatively marked by the grammars, but the final authority on their identification are classification models built from analysis of both TimeBank and large unannotated corpora. *Features* for these models are derived, as we shall see below, from common strategies for exploiting local context, as well as from mining the results—both mark-up and configurational—of the FS grammar cascading.

5   Classification Models for Temporal Analysis

The classification framework we adopt for this work is based on a principle of *empirical risk minimization*. In particular, we use a *linear classifier*, which makes classification decisions by thresholding inner products of feature vectors and weight vectors. It learns weight vectors by minimizing classification errors (*empirical risk*) on annotated training data.

There are good reasons to use linear classifiers; an especially good one is that they allow for easy experimentation with various types of features, without making any model assumptions. This is particularly important in an investigation like ours, where we do not know *a priori* what kinds of features and feature sets would turn out to be most productive.

For our experiments (Section 6), we use the *Robust Risk Minimization* (RRM) classifier [15], a linear classifier, which has independently been shown useful for a number of text analysis tasks such as syntactic chunking [15], named entity chunking [16,17,18], and part-of-speech tagging [19].

In marked contrast to generative models, where assumptions about features are tightly coupled with algorithms, RRM—as is the case with discriminative analysis—enjoys clear separation of feature representation from the underlying algorithms for training and classification. This facilitates experimentation with different feature representations, since the separation between these and the algorithms which manipulate them does not require that the algorithms change. We show how choice of features affects performance in Section 6.

To use classifiers, one needs to design *feature vector representation* for the objects to be classified. This entails selection of some predictive attributes of the objects (in effect promoting these to the status of *features*) and definition of mappings between vector dimensions and those attributes (*feature mapping*). Before we describe (later in this section) the essence of our feature design for EVENT and TLINK recognition,[8] we briefly outline word profiling as the enabling technique for counteracting the paucity of training data in TimeBank.

5.1   Word Profiling for Exploitation of Unannotated Corpora

In general, classification learning requires substantial amount of labeled data for training—considerably more than what TimeBank offers (Section 2). This

---

[8] We do not discuss SIGNAL recognition here, as the `signal` tag itself contributes nothing to EVENT or TLINK recognition, beyond what is captured by a lexical feature over the temporal connective, independent of whether it is tagged as SIGNAL or not.

characteristic of size is potentially a limiting factor in supervised machine learning approaches. We, however, seek to improve performance by exploiting unannotated corpora, which have the natural advantages of being sizable, and freely
available. We use a *word profiling* technique, developed specially for the purposes of exploiting a large unannotated corpus for tagging/chunking tasks [19].
Word profiling identifies, extracts, and manipulates information that charac-
terizes words from unannotated corpora; it does this, in essence, by collecting
and compressing feature frequencies from the corpus, a process which maps the
commonly used feature vectors to frequency-encoded context vectors.

More precisely, word profiling turns co-occurrence counts of words and features (within certain syntactic configurations: *e.g.* 'next word', 'within a phrase',
'head of subject', *etc*) into new feature vectors. Note that this requires pre-
analysis of the unannotated corpus. For example, observing—in that corpus—
that nouns like *"extinction"* and *"explosion"* are often used as syntactic subject
to *"occur"*, and that *"happen"*'s subjects contain *"earthquake"* and *"explosion"*,
helps to predict that *"explosion"*, *"extinction"*, and *"earthquake"* all function like
event nominals. Such a prediction is motivated by the parallel observation about
the preponderance, in the annotated corpus, of event nominals in subject position to *"occur"* and *"happen"* . In Section 6.2, we demonstrate the effectiveness
of word profiling, specifically for EVENT recognition.

5.2   EVENT Recognition as a Classification Problem

Similarly to named entity chunking, we cast the EVENT recognition task as a
problem of sequential labeling of tokens by encoding chunk information into
token tags. For a given *class*, this generates three tags: E:*class* (the last, end,
token of a chunk denoting a mention of *class* type), I:*class* (a token inside of a
chunk), and O (any token outside of any target chunk). The example sequence
below indicates that the two tokens *"very bad"* are spanned by an event-state
annotation.

$$\cdots \ another/\text{O} \ \ very/\text{I:event-state} \ \ bad/\text{E:event-state} \ \ week/\text{O} \ \cdots$$

In this way, the EVENT chunking task becomes a $(2k+1)$-way classification of
tokens where k is the number of EVENT types; this is followed by a Viterbi-style
decoding. (We use the same encoding scheme for SIGNAL recognition.)

The feature representation used for EVENT extraction experiments mimics the
one developed for a comparative study of entity recognition with word profiling
[19]. The features we extract are:

- ○ token, capitalisation, part-of-speech (POS) in 3-token window;
- ○ bi-grams of adjacent words in 5-token window;
- ○ words in the same syntactic chunk;
- ○ head words in 3-chunk window;
- ○ word uni- and bi-grams based on subject-verb-object and preposition-noun
  constructions;
- ○ syntactic chunk types (noun or verb group chunks only);

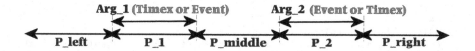

Fig. 4. Partitions for TLINK classifier segmentation

○ token tags in 2-token window to the left;
○ tri-grams of POS, capitalisation, and word ending;
○ tri-grams of POS, capitalisation, and left tag.

5.3   TLINK Recognition as a Classification Problem

TLINK is a relation between events and time expressions which can link two
EVENTs, two TIMEX3's, or an EVENT and a TIMEX3. As we stipulated earlier
(Section 3), presently we focus on TLINKs between events and time expressions.

As a relational link, TLINK does not naturally fit the tagging abstraction
underlying the chunking problem, as outlined above. Instead, we formulate a
classification task as follows. After posting EVENT and TIMEX3 annotations (by
the event classifier and the FS temporal parser, respectively), for each pairing
between an EVENT and a TIMEX3, we ask whether it is a certain type of TLINK.
This defines a $(\ell + 1)$-way classification problem, where ℓ is the number of TLINK
types (before, after, *etc*). The adjustment term '+1' is for the *negative* class, which
indicates that the pair does not have any kind of temporal link relation.

The relation-extraction nature of the task of posting TLINKs requires a different feature representation, capable of encoding the syntactic function of the
relation arguments (EVENTs and TIMEX3's), and some of the larger context of
their mentions. To that end, we consider the following five *partitions* (defined in
terms of tokens): spans of arguments (P_1 or P_2); two tokens to the left/right
of the left/right argument (P_left/P_right); and the tokens between the arguments (P_middle). From each partition, we extract tokens and parts-of-speech
as features (Fig. 4).

We also consider *segments* (*i.e.* syntactic constructions derived by FS analysis:
'when-clause', 'subject', *etc*) in certain relationship to partitions:
○ contained in P_1, P_2, or P_middle;
○ covering P_1 (or P_2) but not overlapping with P_2 (or P_1);
○ occurring to the left of P_1 (or the right of P_2); or
○ covering both P_1 and P_2.
We use uni- and bi-grams of types of these segments as features.

In this feature representation, segments play a crucial role by capturing the
syntactic functions of EVENTs and TIMEX3's, as well as the syntactic relations
between them.

Thus in the example analysis in Fig. 3 (p. 48), svoClause is the smallest segment containing both an EVENT and a TIMEX3, which is indicative of (or at least
does not prohibit) a direct syntactic relation between the two. In the next example (Fig. 5), the TIMEX3 and EVENT chunks are contained in different clauses (a

```
[Snt
 Analysts have complained
 [thatClause that [timex3 third-quarter timex3] corporate earnings
    have n't been very good thatClause]
 [svoClause , but the effect [event hit event] ... svoClause] Snt]
```

Fig. 5. Syntactic configuration discouraging of a TLINK

thatClause and a svoClause, respectively), which structurally prohibits a TLINK relation between the two. Our feature representation is capable of capturing this information via the types of the segments that contain each of EVENT and TIMEX3 without overlapping.

6   Experiments

In line with our current investigation focus (as defined in Section 3), we present here performance results on recognition and typing of TIMEX3, EVENT and TLINK only. Our primary objective here is to report on how effective our analytical strategy is in leveraging the reference nature of the small TimeBank corpus for training classifiers for TimeML. This is the first attempt to build a TimeML-compliant analyser which addresses a more or less full complement of TimeML components; thus there are no comparable results in the literature.

The results (micro-averaged F-measure) reflect experiments with different settings, against the TimeBank corpus, and produced by 5-fold cross validation.

6.1   TIMEX Recognition and Typing

Fig. 6 presents performance results of our TIMEX3 analysis subsystem. Experiments were carried out under different settings. "Span" refers to strict match of both boundaries (the extent) of a TIMEX3 expression; "sloppy" admits time expressions recognised by the FS grammars as long as their right boundary is the same as the reference expression in TimeBank. (One of the observations from the quality analysis of TimeBank reported in [11] is that the corpus is inconsistent with respect to whether some 'left boundary' items—determiners, pre-determiners, and so forth—are marked as a part of the time expression or not; the "sloppy" setting tries to account for this somewhat). As of the time or writing, there are no published results for full TimeML-compliant analysis. We offer here only indirect assessment of our TIMEX3 analysis task, by observing that the numbers for extent marking are not very far from the best systems performance reported at the TERN conference. Of course, given the different definitions of TIMEX2 and TIMEX3, as well as TimeBank's relatively 'cavalier' attitude with respect to TIMEX3's left boundary, the comparison is not very

Task	P	R	F
Span	77.6	86.1	81.7
Span ('sloppy')	85.2	95.2	89.6

	Accuracy
Type (given 'true' span)	81.5

Task	P	R	F
Span + type	64.5	71.6	67.9
Span ('sloppy') + type	70.1	77.8	73.7

Fig. 6. TIMEX3 analysis results, with/without typing. Typing carried out after/ simultaneously with span marking.

meaningful; still, it is indicative of some level of grammar coverage, especially given the incommensurate sizes of the TERN training data and the TimeBank corpus (Section 2).

While TIMEX3 spans are determined by grammars, we use a classifier to type the time expressions. Again, this decision was motivated largely by observing some inconsistencies in type assignment in the corpus, and we felt that, for the purposes of strictly matching the data, machine learning was a more fitting approach to try first (we are yet to compare the typing results presented here with typing by the FS grammars; such a comparison is tied somewhat to getting a better understanding of the quality of annotations in TimeBank). The TIMEX3 typing classifier (second segment of Fig. 6) is configured to use "true" TIMEX3 spans, as per TimeBank, as data points, to which it assigns a category (type) label; thus the table gives a single accuracy measure.

Finally, we report on a joint task, which combines (in sequence) extent marking by FS grammars and type determination as classification process over given spans (this classification task, and features, are defined similarly to the IEO scheme used for EVENT extraction and typing, in without-word-profiling setting; see Section 5.2). In effect, the results here confirm the intuition that imperfect subtasks individually contribute to cumulative degradation of performance.

6.2   EVENT Recognition and Typing

The example analysis in Fig. 3, and the description of features used for the EVENT classification task (Section 5.2) demonstrates how local information and syntactic environment both contribute to the feature generation process. Fig. 7 shows performance results with and without word profiling for exploiting an unannotated corpus.

For the word profiling experiments, we extracted feature co-occurrence counts from 40 million words of 1991 *Wall Street Journal* articles. The proposed event chunks are counted as correct only when both the chunk boundaries and event

features	with typing	w/o typing
basic	61.3	78.6
basic + word-profiling	64.0 (+2.7)	80.3 (+1.7)

Fig. 7. EVENT extraction results, with/without typing. Numbers in parentheses show contribution of word profiling, over using basic features only.

types are correct. 64.0% F-measure is lower than typical performance of, for instance, named entity chunking; this result is indicative of the effects of insufficient training data. On the other hand, a strongly positive indicator here is the fact that word profiling clearly improves performance. In a different setting, when we train the EVENT classifiers without typing, we obtain 80.3% F-measure. This confirms the intuition that the EVENT typing task is inherently complex, and requires more training data.

6.3   TLINK Recognition and Typing

In this experimental setting, we only consider the pairings of EVENT and TIMEX3 which appear within a certain distance in the same sentences (as we will see shortly, this hardly reduces the problem space).[9]

For comparison, we implement the following simple baseline method. Considering the text sequence of EVENTs and TIMEX3's, only 'close' pairs of potential arguments are coupled with TLINKs; EVENT e and TIMEX3 t are close if and only if e is the closest EVENT to t and t is the closest TIMEX3 to e. For all other pairings, no temporal relation is posted. Depending on the 'with-'/'without-typing' setting, the baseline method either types the TLINK as the most populous class in TimeBank, is_included, or simply marks it as 'it exists'.

Results are shown in Fig. 8. Clearly, the detection of temporal relations between events and time expressions requires more than simply coupling the closest pairs within a sentence (as the baseline does). It is also clear that the baseline method performs poorly, especially for pairings over relatively long distances. For instance, it produces 34.9% (in F-measure) when we consider the pairings within 64 tokens without typing. In the same setting, our method produces 74.8% in F-measure, significantly outperforming the baseline.

We compare performance against two types of feature representation: 'basic' and 'basic+FS grammar', which reflect the without- and with-segment-type information obtained by the grammar analysis, respectively. As the positive delta's show, configurational syntactic information can be exploited beneficially by our process. When we focus on the pairings within a 4-tokens window, we achieve 81.8% F-measure without typing of TLINKs, and 58.8% with typing. (The task

---

[9] To evaluate the TLINK classifier alone, we use the EVENT and TIMEX3 annotations in TimeBank. Also, note that the focus on links within a sentence span naturally excludes TLINKs with time expressions in document metadata.

distance (# of TLINKs)	features	with typing	w/o typing
distance ≤ 64 tokens (1370 TLINKs)	baseline	21.8	34.9
	basic	52.1	74.1
	basic+FS	53.1 (+1.0)	74.8 (+0.7)
distance ≤ 16 tokens (1269 TLINKs)	baseline	38.7	61.3
	basic	52.8	75.8
	basic+FS	54.3 (+1.5)	76.5 (+0.7)
distance ≤ 4 tokens (789 TLINKs)	baseline	49.8	76.1
	basic	57.0	80.1
	basic+FS	58.8 (+1.8)	81.8 (+1.7)

Fig. 8. TLINK extraction results, with/without typing. Parentheses show positive contribution of grammar-derived features, over using basic features only. Baseline method posts TLINKs over 'close' pairs of EVENTs and TIMEX3's.

without typing is a binary classification to detect whether the pairing has a TLINK relation or not, regardless of the type.) As the figure shows, the task becomes harder when we consider longer distance pairings. Within a 64 token distance, for instance,, we obtain figures of 74.8% and 53.1%, without and with typing respectively.

While we are moderately successful in detecting the *existence* of temporal relations, the noticeable differences in performance between the task settings with and without typing indicate that we are not as successful in distinguishing one type from another. In particular, the major cause of the relatively low performance of TLINK typing is the difficulty in distinguishing between during and is_included link types.

7   Conclusion

We have used the task of TimeML-compliant parsing to experiment with a specially developed strategy for leveraging minuscule amounts of training data. The strategy synergistically blends finite-state analysis for shallow syntactic parsing with a machine learning technique. The potential for such synergistic approaches to complex analytical problems is clear, especially in situations where reference data—in sufficient quantity, and/or quality—is hard to come by.

This paper highlights two aspects of this blend. We carry out aggressive analysis, by a complex grammar cascade, aiming at considerably more than just partitioning text into chunks: the analysis targets both intrinsic characteristics of temporal expressions, as well as higher-order syntactic configurations used to derive features for a machine learning component. The learning component itself is enhanced by a mechanism specifically designed to counteract paucity in pre-annotated data with the ability to train over unannotated data as well as exploit whatever labeled data is available, no matter how small.

The extreme paucity of the available reference data correlates with the performance results, in particular where the novel components of EVENT and TLINK analysis are targeted, as they appear to fall short of expectations in line with current state-of-the-art information extraction capabilities. Our results are further explained by the inherently noisy nature characteristic of the TimeBank corpus. However, given that the corpus was not designed and populated using rigorous methods for generating training data, our experience is indicative of the effectiveness of a hybrid analytical approach.

Direct comparison of the results reported here with related work is not yet possible. Ours is the first systematic attempt at TimeML-compliant analysis, aiming at a more or less full complement of TimeML components: thus there are no comparable results in the literature.

Mani et al. [20] discuss some pioneering work in linking events with times, and ordering events, suggestive of productive strategies for posting (some) TLINK information. However, the nature of these efforts is such that differences in premises, representation, and focus make a direct performance comparison impossible. Furthermore, the work pre-dates TimeML, and cannot be conveniently mapped to TimeBank data; this, in effect, precludes a quantitative comparison with our work. Most recently, the TARSQI project has been developing strategies and heuristics for particular subsets of TimeML components [21]; again, there is no basis for direct comparison, as only partial overlap exists between the phenomena and attributes targeted by that work and ours (but see [11] for some in-depth analysis of complementary analytic strategies). For this reason, as well as because TARSQI does not explicitly focus on investigating the utility of TimeBank as a training resource, it is not constructive to attempt comparative assessment.

One thing our work makes especially clear is that, given the ability to use unannotated corpora in conjunction with TimeBank to develop a more accurate and felicitous TimeML models, even small improvements to the corpus would significantly boost performance. The corpus would benefit substantially from the application of rigorous methodology for compiling training data. Even a relatively minor effort of cleaning up the existing data would improve performance: this is confirmed by considering the results presented in Section 6 and the corpus characteristics highlighted in Section 2.

A cleanup operation—largely focused on fixing both the errors of omission and commission in the original TimeBank—has now been carried out: TimeBank Version 1.2 represents a considerable improvement over TimeBank 1.1, with respect to largely removing the noise in the first release [11]. TimeBank 1.2 is available through the offices of Linguistic Data Consortium. Future work, of further use to the community, would be an effort to create a larger TimeBank which—by virtue of the systematic methods of developing an annotated corpus within an established set of annotation guidelines—will truly become the widely usable reference resource envisaged from the outset of the TimeML definition and by more recent standardisation efforts [22].

References

1. Pustejovsky, J., Castaño, J., Ingria, R., Saurí, R., Gaizauskas, R., Setzer, A., Katz, G., Radev, D.: TimeML: Robust specification of event and temporal expressions in text. In: AAAI Spring Symposium on New Directions in Question-Answering (Working Papers), Stanford, CA, pp. 28–34 (2003)
2. Hobbs, J., Pan, F.: An ontology of time for the semantic web. TALIP Special Issue on Spatial and Temporal Information Processing 3(1), 66–85 (2004)
3. Saurí, R., Littman, J., Knippen, B., Gaizauskas, R., Setzer, A., Pustejovsky, J.: TimeML annotation guidelines. Technical report, TERQAS Workshop (2005), Version 1.4, (date of citation: February 02, 2006)
   http://timeml.org/site/publications/timeMLdocs/AnnGuide_1.2.1.pdf
4. Advanced Research Projects Agency: In: Proceedings of the Sixth Message Understanding Conference (muc-6), Advanced Research Projects Agency, Software and Intelligent Systems Technology Office (1995)
5. Advanced Research Projects Agency: In: Proceedings of the Seventh Message Understanding Conference (muc-7), Advanced Research Projects Agency, Software and Intelligent Systems Technology Office (1998)
6. Boguraev, B., Ide, N., Meyers, A., Nariyama, S., Stede, M., Wiebe, J., Wilcock, G.: Linguistic Annotation Workshop (the LAW); ACL-2007, Prague, The Czech Republic, Association for Computational Linguistics (June 2007)
7. Fikes, R., Jenkins, J., Frank, G.: JTP: A system architecture and component library for hybrid reasoning. Technical Report KSL-03-01, Knowledge Systems Laboratory, Stanford University (2003)
8. Han, B., Lavie, A.: A framework for resolution of time in natural language. TALIP Special Issue on Spatial and Temporal Information Processing 3(1), 11–35 (2004)
9. Hobbs, J., Pustejovsky, J.: Annotating and reasoning about time and events. In: AAAI Spring Symposium on Logical Formalizations of Commonsense Reasoning, Stanford, CA (March 2004)
10. Pustejovsky, J., Hanks, P., Saurí, R., See, A., Gaizauskas, R., Setzer, A., Radev, D., Sundheim, B., Day, D., Ferro, L., Lazo, M.: The Timebank corpus. In: McEnery, T. (ed.) Corpus Linguistics, Lancaster, pp. 647–656 (2003)
11. Boguraev, B., Pustejovsky, J., Ando, R., Verhagen, M.: Evolution of TimeBank as a community resource for TimeML parsing. Language Resources and Evaluation (Forthcoming 2007)
12. DARPA TIDES (Translingual Information Detection, Extraction and Summarization): The TERN evaluation plan; time expression recognition and normalization. In: Working papers, TERN Evaluation Workshop (2004), (date of citation: July 12, 2005) http://timex2.mitre.org/tern.html
13. Kennedy, C., Boguraev, B.: Anaphora for everyone: Pronominal anaphora resolution without a parser. In: Proceedings of COLING 1996. 16th International Conference on Computational Linguistics, Copenhagen, DK (1996)
14. Schilder, F., Habel, C.: Temporal information extraction for temporal QA. In: AAAI Spring Symposium on New Directions in Question-Answering (Working Papers), Stanford, CA, pp. 35–44 (2003)
15. Zhang, T., Damerau, F., Johnson, D.E.: Text chunking based on a generalization of Winnow. Journal of Machine Learning Research 2, 615–637 (2002)
16. Florian, R., Ittycheriah, A., Jing, H., Zhang, T.: Named entity recognition through classifier combination. In: Proceedings of CoNLL-2003 (2003)

17. Zhang, T., Johnson, D.E.: A robust risk minimization based named entity recognition system. In: Proceedings of CoNLL-2003, pp. 204–207 (2003)
18. Florian, R., Hassan, H., Jing, H., Kambhatla, N., Luo, X., Nicolov, N., Roukos, S.: A statistical model for multilingual entity detection and tracking. In: Proceedings of HLT-NAACL 2004 (2004)
19. Ando, R.K.: Exploiting unannotated corpora for tagging and chunking. Proceedings of ACL 2004 (2004)
20. Mani, I., Pustejovsky, J., Sundheim, B.: Introduction: special issue on temporal information processing. ACM Transactions Asian Language Information Processing 3(1), 1–10 (2004)
21. Verhagen, M., Mani, I., Sauri, R., Littman, J., Knippen, R., Jang, S.B., Rumshisky, A., Phillips, J., Pustejovsky, J.: Automating temporal annotation with tarsqi. In: ACL 2005. 43rd Annual Meeting of the Association for Computational Linguistics, Ann Arbor, Michigan, (Poster/Demo) (2005)
22. Lee, K., Pustejovsky, J., Boguraev, B.: Towards an international standard for annotating temporal information. In: Third International Conference on Terminology, Standardization and Technology Transfer, Beijing, China, ISO TC/37 and SC (August (2006)

Event Extraction and Temporal Reasoning in Legal Documents

Frank Schilder

R&D, Thomson Corp.
610 Opperman Drive, Eagan 55123, U.S.A.
Frank.Schilder@Thomson.com

Abstract. This paper presents a prototype system that extracts events from the United States Code on U.S. immigration nationality and links these events to temporal constraints, such as in *entered the United States before December 31, 2005*. In addition, the paper provides an overview of what kinds of other temporal information can be found in different types of legal documents. In particular, it discusses how one could do further reasoning with the extracted temporal information for case law and statutes.

1 Introduction

In the recent past, little research has been carried out in legal reasoning looking at formalizing temporal information. This should come in particular as a surprise since case law documents, laws, regulations and legal documents in general are normally filled with temporal information:

(1) On **November 12, 1998**, Illinois State Police Trooper Daniel Gillette stopped defendant on Interstate Route 80 in La Salle County for driving 71 miles per hour in a zone with a posted speed limit of 65 miles per hour.
(2) (...) is an alien who entered the United States on or before **December 31, 1990**, who filed an application for asylum on or before **December 31, 1991**, and who, at **the time of filing such application**, was a national of the Soviet Union, (...)
(3) The primary treating physician shall be responsible for obtaining all of the reports of secondary physicians and shall, unless good cause is shown, within **20 days** of receipt of each report incorporate, or comment upon, the findings and opinions of the other physicians in the primary treating physician's report and submit all of the reports to the claims administrator.
(4) Celltech owns a family of patents called the "Adair" patents and sought to claim royalties from Medimmune under a patent licence dated **19 January 1998**.

Although temporal information is actually ubiquitous in legal text, systems for legal reasoning deal normally only on an 'ad-hoc-basis' with this important phenomenon [1]. With the exception of the special issue of Information & Communications Technology Law in 1998 [1,2,3], there is hardly any research on temporal

F. Schilder et al. (Eds.): Reasoning about Time and Events, LNAI 4795, pp. 59–71, 2007.

information in legal text carried out. A couple of recent attempts focused on the specification of legal text in XML including temporal information [4,5,6]. Apart from these few research projects the extraction of temporal information has not been looked at in the literature. Traditionally, legal reasoning has been the focus of AI-related research, where the content of laws and regulations may, for example, become formalized in the event calculus [7]. Time may play a role within such a formalization, but it has not been the main focus of the formalization apart from a few exceptions.

The aim of this paper is to give a first attempt on how temporal information extraction techniques can be married with formal temporal reasoning approaches. Instead of formalizing legal text, we would like to extract parts of legal text a lawyer may be interested via Information Extraction (IE) methods such as finite state transducers. This first task could be coined event extraction where a lawyer, for example, is interested in a specified event type mentioned in various legal documents such as statutes. As a second task, we would like to capture the temporal constraints that may be associated with these events.

As an example, we can assume that a lawyer would like to find entering the United States events in the United States Code 8 (U.S.C. 8) on U.S. immigration and nationality. However, the number of occurrences of the phrase *entered the United States* is quite high in the entire statute and the lawyer may want to filter out only events that describe the actual circumstances of how her client entered the country. In order to do this, we need to extract events that do not contradict the temporal constraints given by the concrete case. A person may have entered the country on November 16, 2005. A section of the U.S.C. 8 that talks about aliens who entered the country before January 1, 1995 is clearly not relevant and should not be presented to the lawyer for further review.

The rest of the paper is organized as follows. We will first look at temporal information in legal text in general. We discuss different types of legal text and investigate what kind of temporal information they can contain and after reviewing how this information could be automatically extracted, we will present our results of a first study on extraction and reasoning with the temporal information in statutes.

Section 2 contains an overview of different kinds of legal documents and provides a brief introduction on how temporal information and constraints can be important for researching these legal documents. Section 3 focuses on a prototype system for event and temporal information extraction implemented within the UIMA framework. Section 4 concludes and discusses possible avenues of future research.

2   Legal Documents and Temporal Information

Legal documents can be categorized in different ways. For this paper, we make the following distinction for different U.S. legal documents:

- Statutes (issued by the federal government)
- Proclamations, code of Federal Regulations, administrative decisions (issued by the President, Executive Departments and administrative departments (e.g. National Labor Relations Board (NLRB))
- Case law (authorized by trial courts, appellate courts or the supreme courts)
- Transactional documents (written by lawyers)
- Documents used as evidence for a case
- News documents that mention parties or people relevant to a case

There are different ways of how to look at temporal information and legal documents. For one thing we can look at the documents and their creation date or the date when the law described by them takes effect. Legal documents can be ordered along a time line according to these dates. This ordering of documents could be called extrinsic temporal ordering.

Another ordering would be an intrinsic temporal ordering of the events described within the document and placing them onto a time line. This type of temporal extraction is clearly more sophisticated and requires deep NLP processing techniques.

Another way of processing temporal information derived from legal documents is the mining of information about the participating parties mentioned in the document. Based on the creation date, one can derive that a lawyer works for a particular company at that time. A different case may show the same lawyer working for a different company at a latter point in time. Other text types such as news messages about companies, law firms or lawyers may also give information about the current affiliation of the people mentioned in the text. This information could be used to update databases on companies, law firms or lawyers.

All these three dimensions of temporal extraction and reasoning can be found if we look at the normal life cycle of a case. Traditionally, the search for precedent cases is the centerpiece for the American legal system and most often the starting point for the legal researcher. Hence, it is absolutely essential to find precedent cases relevant to the current case that are also not superseded by decisions of a higher court made at a later date. Services such as Keycite™ offer a legal researcher the tool to search the *history* and status of U.S. and state court cases and statutes. In order to ensure accuracy this information is annotated by editors a couple of hours after the decisions have become public.

Apart from this classic case of ordering legal cases according to a time line, there are other applications where the automatic temporal ordering of documents can become crucial for a legal researcher. In the following, we will look at two different kinds of legal text in more detail: legal narratives and statutes. The occurrence of temporal expressions in another type of legal text (i.e., transactional documents) is discussed in [8]. Here, we first discuss fact-based narratives in case law which are most similar to news messages, because they mention mainly actual events that are linked to temporal expressions. Second, we investigate what kind of temporal expressions can be found in statutes. They are concerned with normative legal concepts rather than with concrete events. Consequently, event types are described that are linked to temporal expressions. We found a higher number of durations than is normally the case in news messages.

2.1 Legal Narratives in Case Law

Narrative language describing the facts of the case most often contains temporal expressions. At the beginning of a case the judge normally describes the facts and the reasoning that follows should be based on the relevant laws, statutes or regulations relevant to these facts.

(5) On November 12, 1998, Illinois State Police Trooper Daniel Gillette stopped defendant on Interstate Route 80 in La Salle County for driving 71 miles per hour in a zone with a posted speed limit of 65 miles per hour. Trooper Gillette radioed the police dispatcher that he was making the traffic stop.

Such narratives are very similar to news messages and an off-the-shelf temporal tagger could extract temporal expressions reasonably well from this type of text. In addition research focusing on temporal information derived from narratives [9] could be leveragesd for deriving a formal representation of the chain of events. Having derived the temporal constraints on the event described in the case, searches could be carried out that contain temporal constraints. A query such as "*Banana /s slip /before fall*" would return only cases where a *slipping* event occurred before a *falling* event. Note that this is a (temporal) relation between events and not sentences.

2.2 Temporal Restrictions in Statutes or Regulations

Statutes and regulations contain several different types of temporal expressions. In contrast to the fact-based narratives one finds in case law, they often contain periods of time (e.g. *30 days*) or sets of times (e.g. *every year*). These two types of temporal expressions are used to add time constraints to event types rather than to an actual event, as this is the case in news messages or the facts sections of a case.

(6) ATTORNEY GENERAL OPTION TO ELECT TO APPLY NEW PROCEDURES.- In a case described in paragraph (1) in which an evidentiary hearing under section 236 or 242 and 242B of the Immigration and Nationality Act has not commenced as of the title III-A effective date, the Attorney General may elect to proceed under chapter 4 of title II of such Act (as amended by this subtitle). The Attorney General shall provide notice of such election to the alien involved **not later than 30 days** before the date any evidentiary hearing is commenced. If the Attorney General makes such election, the notice of hearing provided to the alien under section 235 or 242(a) of such Act shall be valid as if provided under section 239 of such Act (as amended by this subtitle) to confer jurisdiction on the immigration judge.

The anchor for the duration in (6) is found in the date an evidentiary hearing is commenced. It is important to note that the link between the temporal expression and this event is conditional. Only if such an evidentiary hearing exists does the 30-days restriction apply.

Statutes may also contain date expressions. These can be linked to an actual event, as for an effective date (or termination date) in (e.g. (7)). But mostly, even these date expressions are linked to an event type as a temporal constraint, as in (8).

(7) Amendment by Pub. L. 99177 effective **Dec. 12, 1985**, and applicable with respect to fiscal years beginning after Sept. 30, 1985, but with subsec. (c) to expire **Sept. 30, 2002**, see section 275(a)(1), (b) of Pub. L. 99177, as amended, set out as an Effective and Termination Dates note under section 900 of Title 2, The Congress.

(8)) (...) is an alien who entered the United States on or before **December 31, 1990**, who filed an application for asylum on or before **December 31, 1991**, and who, at **the time of filing such application**, was a national of the Soviet Union, Russia, any republic of the former Soviet Union, Latvia, Estonia, Lithuania, Poland, Czechoslovakia, Romania, Hungary, Bulgaria, Albania, East Germany, Yugoslavia, or any state of the former Yugoslavia;

In a preliminary study of the United State Code we investigated the performance of an off-the-shelf temporal tagger (i.e. TempEx by [10]) on a small test set drawn from the United States Code by hand-annotating this test set with respect to the links between temporal expressions and events or event types according to the TimeML specification [11]. TimeML is a specification language for the annotation of events and temporal expressions. Events, for example, are annotated by EVENT tags, temporal expressions by TIMEX tags and the relation between events and temporal expressions are indicated by TLINK.

First we ran the TempEx tagger and computed precision and recall for a randomly selected set of 26 statute sections extracted from the 8th United States Code on Aliens and Nationality. Of the 64 temporal expressions in the sampled sections, the temporal tagger identified 24. Of these, four, contained incorrect date attributions. Results on this test are shown in table 1. Take into consideration that the Tempex tagger was written for news messages and that such a test can only be seen as a baseline for temporal taggers that are more fine-tuned for legal language in statutes or regulations.

Then we hand-annotated all temporal expressions in these 26 sections according to the subordinated link and temporal link between the temporal expression

Table 1. Temporal tagging accuracy

	correct	occurrences	percent
Precision	20	24	83.33%
Recall	20	64	31.25%

Table 2. Distribution of temporal expressions in subset of U.S.C. 8

	ET			AE
	Period	Set	Date	Date
	22	26	11	5
		59		5
total		64		

and the event (type). We distinguished the following two categories: (a) an event type describes an event that is not necessarily anchored on the time line (e.g., emphan alien who entered the United States before January 1st, 1999). Formally, the event variable can be bound by a universal quantifier (i.e., $\forall e \; enter(e) \rightarrow \tau(e) < 1999 - 01 - 01$).

In addition to the distinction between and event type and an actual event, we investigated which temporal modifiers (e.g., frequencies or dates) co-occur with the events. We found the following combinations of event type/actual event and temporal modifiers:[1]

ET-Period Event type linked to a period describes an event that has to happen within a definite period of time (e.g., *needs to file within 20 days*)

ET-Set Event type linked to a set of times (or a frequency) indicates an event frequency (e.g., *must not enter more than two times*)

ET-Date Event type linked to a date specifies when an event should (or should not have) happened in order to meet some condition (e.g., *an alien who entered the United States before January 1st, 1999*)

AE-Date An actual event linked to a date (e.g., *John Smith entered the country on January 5th, 2005*)

The event type definition can be seen as similar to the event ordering definition provided by [12] (i.e., *establishing the relative position of two events in time*).

The results of our preliminary study of the distribution of different types of links between temporal expressions and event (types) can be found in figure 2. From the distribution of these different link types one can conclude that temporal expressions in statutes serve a different function than in news messages or in the facts sections of cases. Statutes define event types that can be restricted by temporal constraints. A set of people may be defined by their actions within a certain time frame in addition to other conditions that have to hold (e.g. (8)). Such conditional definitions do not occur that often in factive text.

Nevertheless, the TimeML specification allows for such a link via an SLINK [11]:

(9) Bush held out the prospect of more aide$_{e_{248}}$ to Jordan if$_{s_{1298}}$ it cooperates$_{e_{249}}$ with the trade embargo.

---

[1] We did not find any periods or sets of times linked to actual events (e.g. *John wrote the note within 2 minutes*).

```
<SLINK eventInstanceID="e248" subordinatedEventInstance="e249"
signalID="s1298" relType="CONDITIONAL"/>
```

Important signals for conditional SLINKs are conjunctions *when* or *if*, as described in the TimeML annotation guide. TimeBank 1.2[2] contains indeed 45 SLINKS which are almost exclusively signaled by *if*-constructions (i.e., 39/45). Those signals, however, are not found in statutes. Instead these temporal expressions are often used within a modal context (cf. *The Attorney General shall provide notice of such election to the alien involved **not later than 30 days***).

Extracting these links can be useful for the shallow processing of statutes where conditions including temporal ones are extracted and a matching algorithm could filter those statutes or regulations relevant to a given case (e.g. *former citizen of East Germany entered the United States om November 11th, 1990 and filed an application for asylum 20 days after he entered the country* fulfills all conditions stated in (8)).

Another important temporal dimension one encounters with this type of document is the history of the statute. Arnold-Moore describes a system that keeps track of the amendments that were added to a statutes of regulation. This system is currently being used for legislations in Tasmania.[3]

3 Reasoning with Temporal Information and Event Types

As a case study, we extracted sentences describing entering the US event type descriptions from the U.S.C. 8. The goal of this study was to show how temporal information extracted for event types can be used to match description of real cases such as in (10)).

(10) Juan Anibal Aguirre-Aguirre entered the United States without inspection in 1993 and applied for asylum and withholding of deportation.

3.1 System Description

We implemented the prototype as several UIMA[4] components that produced Prolog clauses as output representing the syntactic structure, the named entities and the date and time information. Given these clauses, a temporal reasoner is able to determine whether sentences from the U.S.C. 8 match with given descriptions of concrete cases. Our prototype consists of the following analysis engines:

– Tokenizer and sentence splitter
– Syntactic analysis

---

[2] http://timeml.org/site/timebank/timebank.html
[3] http://www.thelaw.tas.gov.au/index.w3p
[4] http://www.alphaworks.ibm.com/tech/uima

- Named entity tagging
- Time and date extraction
- Event extraction

For the first three engines, we used openNLP, an open source tool for POS tagging, sentence splitting and shallow parsing.[5] We developed our own DateTime annotator that also derives the meaning of the temporal prepositional phrase such as *before December 31, 1990*. The temporal reasoning part was carried out on the generated Prolog clauses.

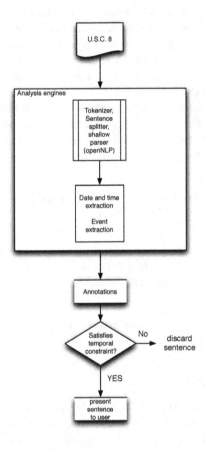

Fig. 1. System overview

Time and date extraction. The annotation of temporal expressions and their temporal relations carried out by a simple Analysis Engine focussing in particular on the temporal expressions in entering events. First of all, each temporal expression has a representation within an ISO 8601-like representation [13]. Secondly,

---

[5] http://opennlp.sourceforge.net/

we added temporal prepositions to the derived temporal information and finally, used some basic temporal functions (e.g., adding and subtracting times).

Temporal representation format. We use ISO 8601 with some extensions that capture in particular time periods with definite and indefinite beginning or ending point. Generally, an ISO expression can either be anchored or unanchored. The anchored expression or timestamp (TS) contains the date and time information of the following form: YYYY-MM-DDTHH:MM:SS. The string 2007-07-07TXX:XX:XX, for example, indicates July 7th, 2007. The granularity level is day and the time information is underspecified, as indicated by the Xs. For brevity, the underspecified time information is often omitted.

An unanchored expression is a duration and is schematically represented as PNG, where N is either a number or X, and G is the abbreviation for the given granularity level. The duration two months, for instance is encoded as P2M and the unspecified temporal expression weeks is represented as PXW.

Durations can also be anchored if a timestamp is supplied by the context. In such a case the unanchored duration PNG is turned into an anchored time stamp. For example the period of three days (i.e. P3D), as in *the next three days*, can be anchored with a time stamp (e.g., *July 7, 2007*) and turned into 2007-07-07P3D which represents the closed time interval starting with July 7, 2007 and ending with July 9, 2007. Conversely, we can add a time stamp at the ending of this period which results in the expression 2007-07-07PB3D referring to a three day interval with the last day being July 6, 2007. Note that TSPB intervals are open with respect to the time stamp, whereas TSP intervals with the time stamp at the left end of the period includes the time stamp.

Temporal relations. Our DateTime tagger also tags temporal prepositions. Note that temporal prepositions contribute to the temporal meaning of temporal expressions [14]. For the prototype, we adopted the specification of temporal prepositions, as described in [13]. The preposition *before*, for instance, is defined as a function that takes an anchored TS and gives back an interval TSPBXG, where G is the granularity of TS.

In order to relate the temporal expression to the event time of the event (i.e., $\tau(e)$), a subset relation between the interval described by the temporal PP t_{PP} and the event time is stipulated (i.e. $\tau(e) \subset t_{PP}$). This representation is equivalent to an alternative representation where the temporal preposition introduces the temporal relation (e.g., $\tau(e) < t_{NP}$), where t_{NP} is the time denoted by the NP in the temporal PP (e.g., *Sunday* in *before Sunday*) provided that t_{PP} is an open interval as defined earlier (e.g., 2007-07-07PBXD).

In a news context, however, the event most likely occurred in a couple of days before (or after) the anchor date. The sentence *He left after July 7, 2007* in a news context, for example, describes a situation where the *leaving* event occurred a couple of days after July 7, 2007. Consequently, the time stamp 2007-07-07PBXD may be further constrained by $X < 5$.

In the legal context, however, time periods can sometimes be fully underspecified with respect to the beginning and ending point of the related time interval.

An expression such as *before December 1st, 1999* for example, refers to a period with no definite beginning point.[6] However, the meaning of the temporal expression is still 1999-12-01TPBXD indicating a granularity level of days. The reasoning component has to consider which interpretation of X is preferred. A legal description prefers the interpretation with an open time interval, whereas a news context indicates a limited number of days before the anchor time.

Temporal functions. In order to compute the correct temporal information, a couple of temporal functions need to be employed. Such functions include, for example, adding or subtracting date and time information. For the prototype, we used some basic temporal functions from the package joda-Time[7] and implemented additional ones. Most importantly, we need to address the question of conjunctions in temporal PPs, such as in *on or before January 1, 2000*. Given a disjunction, we compute two temporal PPS (i.e., *on January 1, 2000* and *before January 1, 2000*). The time stamps are translated into their ISO representations, respectively (i.e., 2000-01-01P1D and 2000-01-01PBXD). A function **add** gives back a new timestamp, if it represents a consecutive time span: 2000-01-02PBXD. Similarly, *on or after January 1, 2000* is translated to 2000-01-01PXD.

Event extraction. The end result of this pipeline is a database of sentences that described an entering the United States event and a temporal constraint. Given a sentence that describes a concrete case, we are now able to match the actual event against the event type description, such as in (10).

(11)(a) entered the United States prior to January 1, 1972.

The event extraction module is a finite state automaton that checks for the occurrence of the verb *entered* and a subsequent occurrence of the NP *the United States*. In addition, the temporal PP is extracted and the meaning is computed. In example (11), the following clause can be derived:

```
clause(E, enter, 'United States', '1972-01-01PBXD', 'USC8', 1259).
```

This clause contains the temporal constraint that the entering event had to occur within the time frame of before January 1st, 1972 as well as the pointer to the section in the U.S.C. 8. After mining the code for entering events as well as the temporal constraints, a knowledge base of these event types can be established. Checking this data base with a concrete case involving a person who entered the United States at a certain date can easily be done via the following query:

```
clause(E,enter,'United States',T, Codes, Section),
temp_subset('1970-04-04', T).
```

---

[6] One could, however, argue that the earliest point in time would be the creation of the United States Codes.

[7] http://joda-time.sourceforge.net/

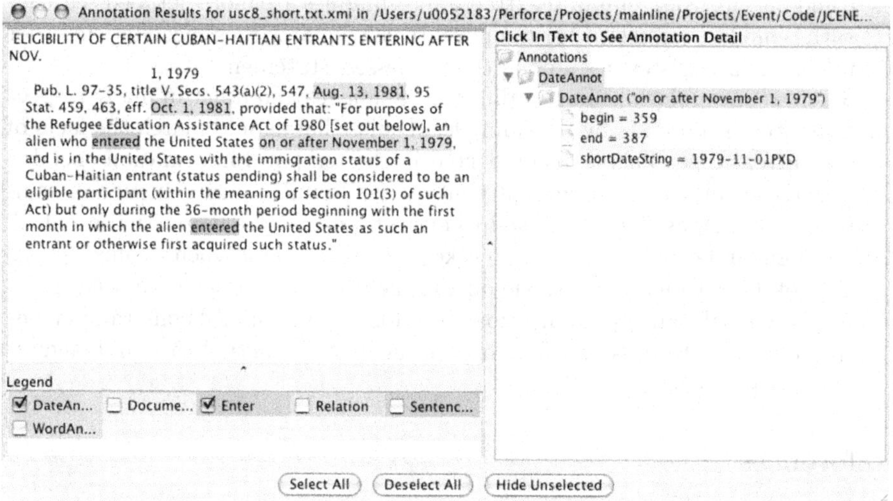

ELIGIBILITY OF CERTAIN CUBAN-HAITIAN ENTRANTS ENTERING AFTER NOV.

1, 1979

Pub. L. 97-35, title V, Secs. 543(a)(2), 547, Aug. 13, 1981, 95 Stat. 459, 463, eff. Oct. 1, 1981, provided that: "For purposes of the Refugee Education Assistance Act of 1980 [set out below], an alien who entered the United States on or after November 1, 1979, and is in the United States with the immigration status of a Cuban-Haitian entrant (status pending) shall be considered to be an eligible participant (within the meaning of section 101(3) of such Act) but only during the 36-month period beginning with the first month in which the alien entered the United States as such an entrant or otherwise first acquired such status."

Click In Text to See Annotation Detail

Annotations
▼ DateAnnot
  ▼ DateAnnot ("on or after November 1, 1979")
      begin = 359
      end = 387
      shortDateString = 1979-11-01PXD

Legend
☑ DateAn... ☐ Docume... ☑ Enter ☐ Relation ☐ Sentenc...
☐ WordAn...

(Select All) (Deselect All) (Hide Unselected)

Fig. 2. An entering event and its temporal constraint

3.2 Extensions

Future extensions include the computation of further temporal constraints and the representation of other constraints, such as citizenship of the alien or of the filing date for asylum.

Another important extension in order to make this approach usable for a lawyer is to compute the validity of the statute section. Our system runs on the raw U.S.C. text and does not consider which parts may not be valid any more because of later amendments of the statute. Documents that take the amendments into account, however, can easily be made available to the lawyer based on editorial enhancements.

4   Conclusions

This paper reports on work-in-progress on event extraction and temporal information extraction and reasoning techniques for legal documents. More specifically, we presented first results in modeling an extraction and reasoning tool for *entering the United States* events in the U.S.C. 8.

In general, we find that most legal text contain many temporal expressions that could be mined and used for automatic reasoning systems for a variety of purposes that may be interesting for the legal practitioner:

- Legal narratives in case law are similar to news messages and off-the-shelf temporal taggers should provide a good coverage with respect to extracting temporal expressions. In addition, the narrative structure should give additional clues for ordering the events of the current case. Applications that could benefit from temporal extraction techniques are more detailed searches with temporal connectors or temporal reasoning of witness accounts in order to detect inconsistencies among the witnesses' statements.

- Statutes or regulations have a different languages and differ in many respect from other legal texts by providing legal rules that should match the facts of the current case. This is also reflected in the temporal information encoded into these rules. In a preliminary study, we found a large amount of temporal expressions that are linked to event types rather than actual event. A temporal and event tagger has to take this into account when applied to this kind of data. Consequently, the off-the-shelf temporal tagger we used had a very low recall. Future applications could use the temporal constraints mentioned in the statutes and match them against the actual case and suggest relevant passages.

References

1. Vila, L., Yoshino, H.: Time in automated legal reasoning. Information and Communications Technology Law 7, 173–197 (1998)
2. Knight, B., Ma, J., Nissan, E.: Representing temporal knowledge in legal discourse. Law, Computers, and Artificial Intelligence / Information and Communications Technology Law 7(3), 199–211 (1998)
3. Farook, D.Y., Nissan, E.: Temporal structure and enablement representation for mutual wills: Law, Computers, and Artificial Intelligence / Information and Communications Technology Law 7(3), 243–268 (1998)
4. Arnold-Moore, T.: About time: legislation's forgotten dimension. In: Proceedings of the 3rd AustLII Law via the Internet Conference 2001, Sydney, Australia (November 2001)
5. Arnold-Moore, T.: Point in time publication for legislation (xml and legislation). In: Proceedings ot the 6th Conference on Computerisation of Law via the Internet, Paris, France (December 2004)
6. Grandi, F., Mandreoli, F., Tiberio, P., Bergonzini, M.: A temporal data model and system architecture for the management of normative texts (extended abstract). In: Proceedings of SEBD 2003 - Natl'. Conf. on Advanced Database Systems, Cetraro, Italy, pp. 169–178 (June 2003)
7. Kowalski, R., Sergot, M.: A logic-based calculus of events. New Gen. Comput. 4(1), 67–95 (1986)
8. Schilder, F., McCulloh, A.: Temporal information extraction from legal documents. In: Katz, G., Pustejovsky, J., Schilder, F. (eds.) Annotating, Extracting and Reasoning about Time and Events. Dagstuhl Seminar Proceedings, Internationales Begegnungs- und Forschungszentrum fuer Informatik (IBFI), Schloss Dagstuhl, Germany. Dagstuhl Seminar Proceedings, vol. 05151 (2005), (date of citation: January 1, 2005), http://drops.dagstuhl.de/opus/volltexte/2005/313
9. Mani, I., Pustejovsky, J.: Temporal discourse models for narrative structure. In: Webber, B., Byron, D.K. (eds.) Proceedings of the ACL, Workshop on Discourse Annotation, Barcelona, Spain, Association for Computational Linguistics, pp. 57–64 (July 2004)
10. Mani, I., Wilson, G.: Robust temporal processing of news. In: ACL 2000. Proceedings of the 38th Annual Meeting of the Association for Computational Linguistics, Hong Kong, pp. 69–76 (June 2000)

11. Pustejovsky, J., Ingria, B., Sauri, R., Castano, J., Littman, J., Gaizauskas, R., Setzer, A., Katz, G., Mani, I.: The specification language TimeML. In: Mani, I., Pustejovsky, J., Gaizauskas, R. (eds.) The Language of Time: A Reader, Oxford University Press, Oxford (February 2005)
12. Pustejovsky, J., Knippen, R., Littman, J., Saurí, R.: Temporal and event information in natural language text. Computers and the Humanities 39(2-3), 123–164 (2005)
13. Schilder, F.: Extracting meaning from temporal nouns and temporal prepositions. ACM Trans. Asian Lang. Inf. Process. 3(1), 33–50 (2004)
14. Schilder, F., Habel, C.: From temporal expressions to temporal information: Semantic tagging of news messages. In: Proceedings of ACL2001 workshop on temporal and spatial information processing, Toulouse, France, pp. 65–72 (2001)

Computational Treatment of Temporal Notions: The CTTN–System

Hans Jürgen Ohlbach

Institut für Informatik, Universität München
ohlbach@lmu.de

Abstract. The CTTN–system is a computer program which provides advanced processing of temporal notions. The basic data structures of the CTTN–system are time points, crisp and fuzzy time intervals, labelled partitionings of the time line, durations, and calendar systems. The labelled partitionings are used to model periodic temporal notions, quite regular ones like years, months etc., partially regular ones like timetables, but also very irregular ones like, for example, dates of a conference series. These data structures can be used in the temporal specification language GeTS (GeoTemporal Specifications). GeTS is a functional specification and programming language with a number of built-in constructs for specifying customised temporal notions.

CTTN is implemented as a Web server and as a C++ library. This paper gives a short overview over the current state of the system and its components.

1 Introduction

In the CTTN–project we aim at a very detailed modelling of the temporal notions. These are, in particular, time points, crisp and fuzzy temporal intervals together with built-in as well as user definable relations between and operations on these intervals. Furthermore, there is support for various kinds of regular and irregular periodic temporal notions, again built-in ones as well as user definable ones. The possibilities range from very simple ones like seconds or minutes up to complex ones like Easter time or solar eclipses. A special specification and programming language GeTS (GeoTemporal Specifications [10]) allows applications and users to defined their own versions of temporal notions and to do all kinds of computations with them.

CTTN is *not* the implementation of a theoretical temporal logic, but it models the flow of time as it is perceived on our planet. It realizes

the main concepts and operations underlying many temporal notions in natural language.

The key components of the CTTN–system consist of the modules depicted in Figure 1. The Service module at the bottom contains a large variety of application independent functions. The FuTI module (Fuzzy Time Intervals) [9,8] contains the data structures and operations on time time points and crisp and

F. Schilder et al. (Eds.): Reasoning about Time and Events, LNAI 4795, pp. 72–87, 2007.
© Springer-Verlag Berlin Heidelberg 2007

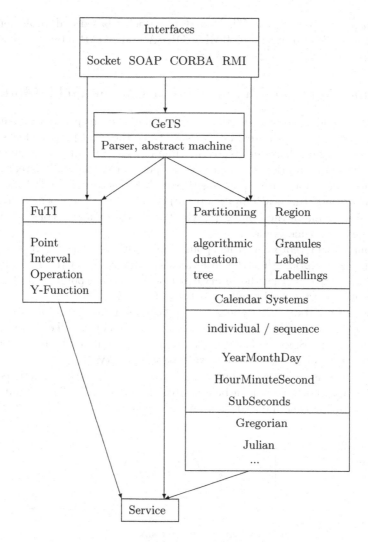

Fig. 1. The CTTN-System

fuzzy time intervals. The largest module is the PartLib module (Partitioning Library). It contains the machinery for specifying and working with periodic temporal notions. Since calendar systems consist of such periodic temporal notions, a module for representing different calendar systems is also part of PartLib.

The GeTS module implements a functional programming language with certain additional constructs for this application area. A flex/bison type parserc and an abstract machine for GeTS has been implemented as part of the CTTN–system. GeTS is the first specification and programming language with such a rich variety of built-in data structures and functions for GeoTemporal notions. In a first case study it has been used to define various versions of fuzzy interval–interval relations [8].

The basic interface to the CTTN system is socket based and implements the CTTN protocol. Prototypes of RMI, CORBA and SOAP interfaces have also been implemented, but not yet fully tested.

2   Time Points and Time Intervals in the FuTI–Module

The flow of time underlying most calendar systems corresponds to a time axis which is isomorphic to the real numbers $\mathbb{R}$. Since the most precise clocks developed so far, atomic clocks, measure the time in discrete units, it is sufficient to restrict the representation of concrete time points to *integers*. Therefore FuTI represents time points with integers, either with 64–bit integers, or with multiple precision integers (this is a compiler option). Within FuTI there is no assumption about the meaning of these integers, whether they are days, seconds, femtoseconds or not even time points[1].

Although FuTI represents time points only with integers, there is still the underlying assumption that the time axis is isomorphic to the real numbers. That means, for example, the interval between the time points 0 and 1 is not empty, but it is set of real numbers between 0 and 1.

The next important data type is that of time intervals. Time intervals can be crisp or fuzzy. With fuzzy intervals one can encode notions like 'around noon' or 'late night' etc. Since fuzzy intervals are more general and more flexible than crisp intervals, FuTI uses fuzzy intervals as basic interval data type.

Fuzzy intervals are usually defined through their membership functions [15,4]. A membership function maps a base set to real numbers between 0 and 1. The base set for fuzzy time intervals is a linear time axis.

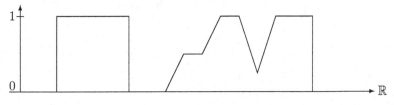

Crisp and Fuzzy Intervals

The fuzzy intervals can also be infinite. For example, the term 'after tonight' may be represented as a fuzzy distribution which rises from fuzzy value 0 at 6 pm until fuzzy value 1 at 8 pm and then remains 1 ad infinitum.

Fuzzy time intervals are realized in the FuTI–module as polygons with integer coordinates. The x-coordinates represent time points and the y-coordinates represent fuzzy values as integers between 0 and a maximum value (the default value is 1000). A normalised fuzzy value between 0 and 1 can then be obtained by dividing the integer y-coordinate by the maximum value. A y-coordinate of 500, for example, represents the normalised fuzzy value 0.5.

---

[1] A special component of FuTI, which was developed for another application allows for the representation of circular intervals like angles between 0 and 360 degrees. In this case the integers represent fractions of angular degrees.

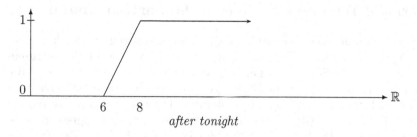

after tonight

If the integers represent hours, one can, for example, represent the interval 'around noon' as the polygon ((11,0) (12,1000) (13,0)). The membership function of the corresponding fuzzy interval starts at 11 o'clock with fuzzy value 0 and then rises linearly to fuzzy value 1 at noon. From there on it falls linearly to fuzzy value 0 at 1 pm.

FuTI provides a large collection of operations on these intervals. There are methods for accessing information about the intervals, the location of various parts of an interval, its size (which is the integral over the membership function), its components etc. There are methods for transforming the intervals, for example, hull computations, there are integration functions, fuzzification functions etc. There are also very general unary and binary transformation functions which can be parameterised with functions operating on the fuzzy values. All the set operations on fuzzy intervals, for example, are realized as transformations with functions on the fuzzy values. The transformations of the fuzzy membership functions need not be linear, i.e. they may transform straight lines into curved lines. The FuTI–module contains for these cases an approximation algorithm which approximates curved lines by polygons.

Example 1 (Birthday Party Time). This example illustrates some of the operations which are possible with the FuTI–module. Consider the statement "the birthday party for took place *from around noon until early evening* of 20/7/2003". The corresponding fuzzy interval could be generated by integrating the fuzzy interval for 'around noon' in positive direction, integrating the fuzzy interval for 'early evening' in negative direction and then intersecting the two integrals. The resulting fuzzy set is:

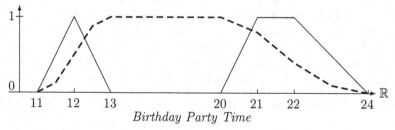

Birthday Party Time

A GeTS specification of this example is given in Example 9.   ∎

3 Periodic Temporal Notions in the PartLib–Module

The PartLib module offers powerful machinery for specifying and working with periodic temporal notions. The basic concept is the concept of the *partitionings of the time axis*. Since most periodic temporal notions, for example, days, yield infinite partitionings of the time axis, PartLib offers different versions of finite representations of these infinite structures. The operations on the infinite structures are turned into operations on the corresponding finite representations.

Partitions can be *labelled*, e.g., with 'Monday', 'Tuesday' etc. Partitionings with labels can be comprised in different ways to different structures. For example, from the day–partitioning and the corresponding labelling one can derive the structure which corresponds to 'all Mondays' or to 'all non-Mondays'. If the labels are organised in a hierarchy, for example, Monday,..,Friday are all 'Workdays' and Saturday and Sunday are 'Weekenddays' one can derive the notion of 'all Workdays'. Since there are a number of further ways to derive new substructures of the time axis from labelled partitionings, all these ways are comprised into the concept of *region structure* (see Sec. 3.5). A region structures is essentially a subset of a partitioning of the time axis. Many operations in the CTTN system work with the more general region structures instead with partitionings.

3.1 Partitionings of the Time Axis

Most basic time units of calendar systems, years, months etc., are essentially partitionings of the time axis. Other periodical temporal notions, for example, semesters, school holidays, sunsets and sunrises etc., can also be modelled as partitionings.

A partitioning of the real numbers $\mathbb{R}$ may be, for example, $(..., [-100, 0[,$ $[0, 100[, [100, 101[, [101, 500[, ...)$. The intervals in the partitionings need not be of the same length (because time units like years are not of the same length either). The intervals can, however, be enumerated by integers (their *coordinates*). For example, we could have the following enumeration

$$... [-100 \; 0[\; [0 \; 100[\; [100 \; 101[\; [101 \; 500[\; ...$$
$$...\quad -1 \qquad 0 \qquad 1 \qquad 2 \qquad ...$$

The enumeration of partitions, i.e. their coordinates, are a very useful means for concrete computations. It turned out, however, that in some cases instead of integer coordinates, certain other structures which are isomorphic to integers are more useful. An example for a structure which is isomorphic to the integers are the paths in an infinite tree. Therefore PartLib has introduced the concept of *Partition Access Specifier (PASp)* as a generalisation of the integer coordinates.

Definition 1 (Partitioning). *A partitioning P of the time axis in PartLib is a sequence*

$$... [t_{-1}, t_0[, [t_0, t_1[, [t_1, t_2[, ...$$

of non-empty half open intervals in $\mathbb{R}$ with integer boundaries such that $t_i < t_{i+1}$ for all i.

The partitioning may be finite at one or both sides, i.e. $]-\infty, t_0[, ..., [t_n, +\infty[$
is allowed.

A Partition Access Specifier Structure *is a set of objects which is isomorphic to the integers.*

A coordinate mapping c *is a bijective mapping between a partitioning and a Partition Access Specifier Structure (or a part of it if the partitioning is finite) such that if partition p is before partition q then $c(p) < c(q)$.* ∎

The choice of half open intervals of the kind $[t_i, t_{i+1}[$ as partitions was arbitrary. It means that, for example, Midnight always belongs to the next day.

3.2 Labelled Partitionings

The partitions in CTTN can be *labelled*. The labels are just names for the partitions like in the following example.

Example 2 (The Labelling of Days). We count the time in seconds beginning with January 1st 1970. This was a Thursday. Therefore we choose as labelling for the day partitioning

$$L \stackrel{\text{def}}{=} Th, Fr, Sa, Su, Mo, Tu, We.$$

The following correspondences are obtained:

$$\begin{array}{llll} time: & \ldots [-86400, 0[& [0, 86400[& [86400, 172800[\ldots \\ coordinate: \ldots & -1 & 0 & 1 & \ldots \\ label: & \ldots & We & Th & Fr & \ldots \end{array}$$

This means, for example, $L(-1) = We$, i.e. December 31 1969 was a Wednesday. ∎

Labels are different to coordinates because different partitions can have the same label (e.g., all Mondays). Labellings can be used for three purposes. The first purpose is to get access to the partitions via their names (labels). One can use these names in various GeTS functions. The second purpose is to associate partitions with further attributes. The labels can, for example, serve as keys into databases. The third purpose is to use the labels for grouping partitions together into *regions*. An example is the set of all Mondays. This is no longer a partitioning of the time axis because there are gaps between the Mondays.

Definition 2 (Labels). *A set of* labels *in PartLib is just an arbitrary finite or infinite set[2]*

A label hierarchy *is a binary relation* $\sqsubseteq$ *which orders the labels in a tree.*

A labelling *of a PartLib partitioning is a possibly partial mapping from the partitions into the set of labels.* ∎

---

[2] Labels are in fact instances of subclasses of a class *Label*.

Since a labelling can be partial, not all partitions need to have labels. As an example, where this makes sense, consider the partitioning of hours and the labelling which associates the label 'working hour' with all hours between 8 am and noon and all hours between 1 pm and 5 pm. The other hours don't have labels. This labelling specifies implicitly the concept of 'working day', the concept of 'lunch time', and the concept of 'after work'. These implicit definitions can be made explicit in PartLib by turning them into *region structures* (see below).

3.3   Specification of Partitionings

Partitionings have a finite representation in PartLib. There are the following representations for partitionings.

Algorithmic Partitionings
This type of partitionings is mainly used for modelling the basic time units of calendar systems, years, months etc. The specification consists of an offset against time point 0, an average length of the partitions, and a correction function which corrects the average length to the actual length.

Example 3 (Basic Time Units for the Gregorian Calendar).
The specification of the basic time units as algorithmic partitionings for the Gregorian Calendar are:

second: average length: 1, offset: 0, correction function: $\lambda(n)0$.

minute: average length: 60, offset: 0, correction function: $\lambda(n)0$.

hour: average length: 3600, offset: 0, correction function: $\lambda(n)0$.

day: average length: 86400, offset: 0, correction function: $\lambda(n) - 3600 \cdot h$ if the day n is during the daylight saving time period, 0 otherwise.
The number h is usually 1 (for 1 hour). Exceptions are, for example, the year 1947 in Germany, where in the night of 1947/5/11 the clock was set forward a second time by 1 hour such that the offset against standard time was 2 hours.

week: average length: 604800, offset -259200[3], correction function: again, this function has to return an offset of $-3600 \cdot h$ for the weeks during the daylight saving time periods.

month: average length: 2592000 (30 days), offset 0, correction function: this function has to deal with the different length of the months and the daylight saving time regulations.

year: average length: 31536000 (365 days), offset 0, correction function: this function has to deal with leap years only. The effects of daylight saving time regulations are averaged out over the year.   ∎

Duration Partitionings
They are specified by an anchor time and a sequence of *'durations'*.

---

[3] This is because the first of January 1970 is Thursday.

For example, I could define 'my weekend' as a *duration partitioning* with anchor time 2004/7/23, 4 pm (Friday July, 23rd, 2004, 4 pm) and durations: ('8 hour + 2 day', '4 day + 16 hour'). The first interval would be labelled 'weekend.

A simpler example is the notion of a semester at a university. In the Munich case, the dates could be: anchor time: October 2000. The durations are: 6 months (with label 'winter semester') and 6 months (with label 'summer semester'). This defines a partitioning with partition 0 starting at the anchor time, and then extending into the past and the future. The first partition in this example is the winter semester 2000/2001.

The units for the duration are in fact region structures, and not just partitionings. Thus, one can, for example, define durations in terms of *granules*. An example is '3.5 working_days + 1.5 weekends'.

smallskip**Date Partitionings**

In this version we provide the boundaries of the partitions by concrete dates. Therefore the partitioning can only cover a finite part of the time line.

An example could be the dates of the Time conferences: 1994/5/4 Time94 1994/5/4 gap 1995/4/26 Time95 1995/4/26 ... 2004/7/1 Time04 2004/7/3.

Since the intervals between two adjacent dates determine durations, date partitionings are in fact special cases of duration partitionings, and this is how they are treated in PartLib.

Intersection Partitionings

They combine two previously defined partitionings by intersecting their partitions. If the two original partitionings are labelled then a new labelling can be computed by means of *mapping rules* for labels.

As an example, suppose there is a partitioning $p1$ representing the lecture course l, say every Wednesday from 10 am until 12 am. There is a second partitioning $p2$ which represents public holidays. $p2$ is labelled with the holiday names (Easter, Christmas etc.) The holiday name labels are all sub-labels in a label hierarchy with top element 'holiday'. The partitioning which represents the lecture time without the public holidays can be generated by intersecting $p1$ and $p2$ with the following mapping rules

$$l * holiday \mapsto gap;$$
$$l * gap \mapsto l;$$
$$gap * holiday \mapsto gap$$

with the extra provision that adjacent partitions without labels are comprised into a single partition. 'gap' stands for the empty label. The following picture illustrates the example.

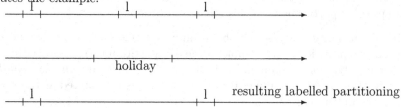

resulting labelled partitioning

Tree Partitionings
This type of specification for partitionings can be used when concrete dates
are involved. Typical examples are bus timetables. A tree partitioning is given
by a *Partition Access Format* (PAF) and a *Partition Access Tree* (PAT). The
PAF determines a kind of calendar to be used for interpreting the nodes in the
PAT [11].

Example 4 (for a Tree Partitioning Specification). A typical PAF is the standard
date format year/week/day/hour/minute/second.
    The following PAT may define a bus schedule.

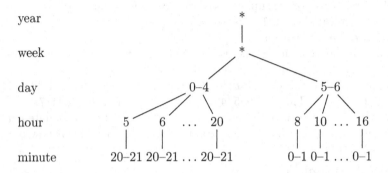

It specifies the following bus schedule: every year, every week, every work day
(0–4), there is a bus at 5:20 – 5.21 (2 minutes stay at the bus stop), 6:20 – 6:21
until 20:20 – 20:21, and at the weekends (days 5,6) there is a bus every hour
from 8 until 16 hours.

    The nodes in the PAT determine an offset from the start of the region given
by the corresponding position in the PAF. There are four different node types:

NumberRange nodes. They specify concrete number sets, for example, 4-6,10-
12 specifies the set $\{4, 5, 6, 10, 11, 12\}$

NumberIterator nodes. They specify iterators like, for example, in a 'for loop'.
The iterator is given by a start value, a step value and a number of iterations.
For example, start = 1, step = 2, iterations = 5 specifies the set $\{1, 3, 5, 7, 9\}$

LabelRange nodes. They specify concrete label sets, for example, March-May,
August specifies the set $\{2, 3, 4, 7\}$ (January is month 0).

LabelIterator nodes. They specify labels by giving a label together with a
number iterator. For example, startLabel = 'L' start = 2, steps = 10, iterations
= 5 The loop starts with the second occurrence of L and then continues 5 times
in steps of 10 partitions with this label, 5 iterations.

    In all four cases it is also possible to interpret the numbers as distances from
the end of a partition. For example, if the day partition is right below the month
partition in the corresponding Partition Access Format, and the backwards flag
is set to true, then the number 0 at the day level is interpreted as the very last
day in the given month.

The specification of a partitioning can be quite complex and require a lot of data. Therefore for each partitioning type, except for algorithmic partitionings, there is a corresponding XML document type for specifying a partitioning. After the CTTN interface has read and parsed such an XML specification one can use them in the same way as the built-in partitionings for calendar systems.

3.4 Leap Seconds

To compensate for the slowing down of the earth's rotation, since 1971 every few years a leap second has been introduced. The last minute in the year where a leap second has been inserted has 61 seconds instead of 60 seconds. This has an effect on all partitionings above the level of seconds. It would be very complicated and error prone to integrate the effect of leap seconds in all these partitionings. As an alternative, this phenomenon is taken care of by separating the reference time into a *global reference time* and a *local reference time*. The global reference time counts the seconds as they are. It knows nothing about leap seconds. The local reference time shrinks the leap seconds to 0 length. That means the last minute in the years where a leap second has been inserted has still 60 seconds in the local reference time. The extra second occurs only in transition to the global reference time. This way the leap second calculations have been concentrated in a single place, the transition between local and global reference time. All other partitioning dependent calculations can ignore leap seconds.

3.5 Region Structures

The labels which can be attached to the partitionings generate a variety of new substructures of the time axis which are no longer partitionings because there can be gaps between the corresponding time intervals. Since periodic temporal notions with gaps are much more frequent than partitionings, the new concept of *region structures* has been introduced.

Region structures are like partitionings, but there are two essential differences

- there are gaps allowed between two neighbouring regions
- there are gaps allowed even within a region. An example is 'working day' from 8 am until 5 pm with a lunch break from 12 am until 1 pm.

CTTN distinguishes the following types of region structures:

PartitioningRegion: each partition is a region. Labels are ignored.

LabelRegion: are determined by a label (possibly within a label hierarchy). For example, the LabelRegion with label 'weekendday' of a day partitioning (with sub-labels Saturday and Sunday below weekendday) would join the days of the weekends into a region. A Saturday is a region, followed by the following Sunday, followed by the following Saturday etc.

LabelBlock: is similar to a LabelRegion. The difference is that neighbouring partitions with the given label form one region. A LabelBlock with labels 'weekendday' (see above) would join Saturdays and Sundays into one single region.

LabellingRegion: declares a whole label sequence as a region. For example, the labelling 'Monday', 'Tuesday', ... 'Sunday' of the day partitioning comprises a whole week into a single region.

GapBlock: A GapBlock comprises all adjacent partitions without labels into one region.

Granule: A granule is a sequence of partitions with the same label possibly interrupted by partitions without label. As an example, consider the hour partitioning where the hours between 8 and 12 and between 13 and 18 hours a labelled 'working hour'. The corresponding granule comprises the working hours into one, in this case non-convex, region. This concept is very much like the concept of granules found in the literature [1].

As soon as a labelling has been attached to a partitioning, all these types of region structures are available as concrete data types, and a common API is available via the superclass 'Region'. Typical examples for the API are methods which move from a given region to the next region, methods which move from a given time point n regions forward or backward (n may be fractional), methods which measure time intervals in terms of region length etc.

3.6   Calendar Systems

A *calendar system* in the CTTN–system is a set of partitionings or region structures, for example the partitionings for seconds, minutes, hours, weeks, months and years, together with some extra data and methods. Dershowitz and Reingold's 'calendrical calculations' are used here [3] for computing the details down to the level of days. In addition PartLib models all the nasty features of real calendar systems, in particular leap seconds and daylight saving time schemes (in a submodule *DLST*). Calendar systems can be arranged in sequences, for example, the sequence consisting of the Julian calendar system until 4th of October 1582 followed by the Gregorian system. Another example of a sequence of calendar systems in PartLib could be a sequence of calendars and time zones a traveller encounters when he travels around the world.

The Calendar submodule in PartLib has predefined general classes for years/ months/days, for hours/minutes/seconds and for sub-seconds. Using these classes it requires very little code to add new calendar systems.

4   The GeTS Language

The PartLib module has, via the XML-interface, mechanisms for integrating user defined periodic temporal notions. Not all temporal notions and computations, however, have to do with periodicies. The GeoTemporal Specification Language GeTS has therefore been added as a general purpose language for working with temporal notions. The design of the GeTS language was influenced by the following considerations:

1. Although the GeTS language has many features of a functional programming language, it is not intended as a general purpose programming language. It is a specification language for temporal notions, however, with a concrete operational semantics.
2. The parser, compiler, and in particular the underlying GeTS abstract machine are not standalone systems. They must be embedded into a host system which provides the data structures and algorithms for time intervals, partitionings etc., and which serves as the interface to the application. GeTS provides a corresponding application programming interface (API).
3. The language should be simple, intuitive, and easy to use. It should not be cluttered with too many features which are mainly necessary for general purpose programming languages.
4. The last aspect, but even more the point before, namely that GeTS is to be integrated into a host system, were the main arguments against an easy solution where GeTS is only a particular module in a functional language like SML or Haskell. The host system was developed in C++ (it could also be Java, but multiple precision integers are more efficient in C++). Linking a C++ host system to an SML or Haskell interpreter for GeTS would be more complicated than developing GeTS in C++ directly. The drawback is that features like sophisticated type inferencing or general purpose data structures like lists or vectors are not available in the current version of GeTS.
5. Developing GeTS from scratch instead of using an existing functional language has also an advantage. One can design the syntax of the language in a way which better reflects the semantics of the language constructs. This makes it easier to understand and use. As an example, the syntax for a time interval constructor is just $[expression_1, expression_2]$.

The GeTS language is a strongly typed functional language with a few imperative constructs. Here we can give only a flavour of the language. The technical details are in [10].

Example 5 (tomorrow). The definition

```
tomorrow = partition(now(),day,1,1)
```

specifies 'tomorrow' as follows: `now()` yields the time point of the current point in time. `day` is the name of the day partitioning. Let i be the coordinate of the day-partition containing `now()`. `partition(now(),day,1,1)` computes the interval $[t_1, t_2[$ where t_1 is the start of the partition with coordinate $i + 1$ and t_2 is the end of the partition with coordinate $i + 1$. Thus, $[t_1, t_2[$ is in fact the interval which corresponds to 'tomorrow'.

In a similar way, we can define

```
this_week(Time t)  = partition(t,week,0,0).
```

The time point `t`, for which the week is to be computed, is now a parameter of the function.   ∎

Example 6 (Christmas). The definition

```
christmas(Time t) =
  dLet year = date(t,Gregorian_month) in
                [time(year|12|25,Gregorian_month),
                 time(year|12|27,Gregorian_month)]
```

specifies Christmas for the year containing the time point t.   ∎

`date(t,Gregorian_month)` computes a date representation for the time point t
in the date format `Gregorian_month` (year/month/day/hour/minute/second).
Only the year is needed. `dLet year = ...` therefore binds only the year to the
integer variable `year`. If, for example, in addition the month is needed one can
write `dLet year|month = date(...`.

`time(year|12|25,Gregorian_month)` computes t_1 = begin of the 25th of
December of this year. `time(year|12|27,Gregorian_month)` computes t_2 =
begin of the 27th of December of this year. The expression $[...,...]$ denotes
the half open interval $[t_1, t_2[$.[4] The result is therefore the half open interval from
the beginning of the 25th of December of this year until the end of the 26th of
December of this year.

Example 7 (Point–Interval Before Relation). The function

```
PIRBefore(Time t, Interval I) =
    if (isEmpty(I) or isInfinite(I,left)) then false
    else (t < point(I,left,support))
```

specifies the standard crisp point–interval 'before' relation in a way which works
also for fuzzy intervals.   ∎

If the interval I is empty or infinite at the left side then `PIRBefore(t,I)` is
`false`, otherwise t must be smaller than the left boundary of the support of I.

Now we define a parameterised fuzzy version of the interval–interval before
relation.

Example 8 (Fuzzy Interval–Interval Before Relation). A fuzzy version of an
interval–interval before relation could be

```
IIRFuzzyBefore(Interval I, Interval J, Interval->Interval B) =
case
  isEmpty(I) or isEmpty(J) or
      isInfinite(I,right) or isInfinite(J,left)      : 0,
  (point(I,right,support) <= point(J,left,support))  : 1,
    isInfinite(I,left) : integrateAsymmetric(intersection(I,J),B(J))
else integrateAsymmetric(I,B(J))
```

     ∎

---

[4] Crisp intervals in CTTN are always half open intervals $[...,...[$. Sequences of such
intervals, for example, sequences of days, can therefore be used to partition a time
period. The syntactic representation of these intervals in GeTS is $[...,...]$ and
not $[...,...[$ because this simplifies the grammar and the parser considerably.

The input are the two intervals I and J and a function B which maps intervals to intervals. B is used to compute for the interval J an interval B(J), which represents the degree of 'beforeness' for the points before J.

The function first checks some trivial cases where I cannot be before J (first clause in the **case** statement), or where I definitely is before J (second clause in the **case** statement). If I is infinite at the left side then $\int (I \cap J)(x) \cdot B(J)(x)dx/$ $|I \cap J|$ is computed to get a degree of 'beforeness', at least for the part where I and J intersect. If I is finite then $\int I(x) \cdot B(J)(x)dx/|I|$ is computed. This averages the degree of a point–interval 'beforeness', which is given by the product $I(x) \cdot B(J)(x)$, over the interval I.

The next example is a parameterised version of an 'Until' operator. It can be used to formalise expressions like 'from around noon until early evening'. The parameters are operators which manipulate the front and back end of the intervals, together with a complement operator.

Example 9 (Until). an 'Until' operator can be defined in GeTS:

```
Until(Interval I, Interval J, Side s1, Side s2,
     (Interval*Interval)->Interval Ints,
     Interval->Interval Ep, Interval->Interval En,
     Interval->Interval C) =
     if (s1 == left) then
        (if (s2 == left) then Ints(Ep(I),C(Ep(J)))
                         else Ints(Ep(I),En(J)))
     else
        (if (s2 == left) then Ints(C(En(I)),C(Ep(J)))
                         else Ints(C(En(I)),En(J)));
```

The birthday party example (Example 1) could be specified using this function:

```
Birthdayparty(I,J)
     = Until(I, J, left, right,
        lambda(Interval K, Interval L) intersection(K,L),
        lambda(Interval K) integrate(K,positive),
        lambda(Interval K) integrate(K,negative),
        lambda(Interval K) complement(K)).
```
∎

5   The Web–Interface

CTTN is a collection of C++ classes and methods which can be used in any other C++ program. There is, however, also a command interface which is realized as a web server. It communicates with a client through a socket. There is a group of commands for uploading application specific definitions of temporal notions in the GeTS language and in the specification language for labelled partitionings. There are also commands for working with instances of these temporal notions, particular time intervals, particular partitionings, particular calendar systems etc.

6   Extensions of the CTTN–System

A number of extensions of the CTTN–system are on the agenda. The most important one is the inclusion of constraint reasoning for 'floating' time intervals. The expression 'two weeks between Christmas and Easter', for example, cannot be represented so far, because the precise location of these two weeks are not known. Here we need to invoke constraints and constraint reasoning. Since the basic intervals are fuzzy intervals, the constraint calculus must also be able to deal with fuzziness. There are some approaches in the direction of fuzzy temporal reasoning [5,14,6] and fuzzy constraint networks [13,7] which might be usable for the CTTN–system. Temporal constraint reasoning without taking fuzziness into account is certainly also very useful and should be integrated into the system [2].

Another extension is a context module. A simple example for context information which is useful for an application of the CTTN–system are the specification of time zones. Time zones are submitted to the current CTTN–system as offsets to GMT time. It would, however, be much more user friendly, if there would be an automatic mapping of countries or regions to time zones.

A third extension is a link to a system which represents *named entities*. The phrase 'after the Olympic games in Rome', for example, can only be analysed if some dates about the Olympic games in Rome are available. We are currently working on a link to the EFGT net, which stores named entities in a three dimensional context of thematic fields, geographic regions and time periods [12].

More details about the CTTN–system are available at the CTTN homepage: http://www.pms.ifi.lmu.de/CTTN.

Acknowledgements

This research has been funded by the European Commission and by the Swiss Federal Office for Education and Science within the 6th Framework Programme project REWERSE number 506779 (cf. http://rewerse.net).

References

1. Bettini, C., Sibi, R.D.: Symbolic representation of user-defined time granularities. Annals of Mathematics and Artificial Intelligence 30, 53–92 (2000)
2. Bry, F., Rieß, F.-A., Spranger, S.: A Reasoner for Calendric and Temporal Data. Forschungsbericht/research report PMS-FB-2005-18, Institute for Informatics, University of Munich (2005)
3. Dershowitz, N., Reingold, E.M.: Calendrical Calculations. Cambridge University Press, Cambridge (1997)
4. Dubois, D., Prade, H. (eds.): Fundamentals of Fuzzy Sets. Kluwer Academic Publishers, Dordrecht (2000)
5. Godo, L., Vila, L.: Possibilistic temporal reasoning based on fuzzy temporal constraints. In: IJCAI 1995. Proceedings of the Fourteenth International Joint Conference on Artificial Intelligence, vol. 2, pp. 1916–1922 (1995)

6. Navarette, I., Cardenas, M.A., Marin, R.: Efficient resolution mechanism for fuzzy temporal constraint logic. In: TIME 2000. Proc. of the Seventh International Workshop on Temporal Representation and Reasoning, pp. 39–46. IEEE Press, Reasoning (2000)
7. Roque Marín, M.A., Viedma, C., Balsa, M., Sanchez, J.L.: Obtaining solutions in fuzzy constraint networks. Int. J. Approx. Reasoning 16(3-4), 261–288 (1997)
8. Ohlbach, H.J.: Relations between fuzzy time intervals. In: Proceedings of 11th International Symposium on Temporal Representation and Reasoning, Tatihoui, Normandie, France, 1st–3rd July 2004, pp. 44–51. IEEE Computer Society Press, Los Alamitos, http://www.pms.ifi.lmu.de/publikationen/#PMS-FB-2004-33
9. Ohlbach, H.J.: Fuzzy time intervals – the FuTI-library. Research Report PMS-FB-2005-26, Inst. für Informatik, LFE PMS, University of Munich (June 2005), URL: http://www.pms.ifi.lmu.de/publikationen/#PMS-FB-2005-26
10. Ohlbach, H.J.: GeTS – a specification language for geo-temporal notions. Research Report PMS-FB-2005-29, Inst. für Informatik, LFE PMS, University of Munich (June 2005), URL:
    http://www.pms.ifi.lmu.de/publikationen/#PMS-FB-2005-29
11. Ohlbach, H.J.: Periodic temporal notions as 'tree partitionings'. Forschungs-bericht/research report PMS-FB-2006-11, Institute for Informatics, University of Munich (2006)
12. Schulz, K.U., Weigel, F.: Systematics and architecture for a resource representing knowledge about named entities. In: Bry, J.M.F., Henze, N. (eds.) Principles and Practice of Semantic Web Reasoning, pp. 189–208. Springer, Berlin (2003)
13. Vila, L., Godo, L.: On fuzzy temporal constraint networks. Mathware and Soft Computing 3, 315–334 (1994)
14. Vila, L., Godo, L.: Query-answering in fuzzy temporal constraint networks. In: Mellish, C.S. (ed.) FUZZ-IEEE 1995. IEEE International Conference on Fuzzy Systems Yokohama, vol. 1, pp. 43–48. IEEE, Los Alamitos (1995)
15. Zadeh, L.A.: Fuzzy sets. Information & Control 8, 338–353 (1965)

Towards a Denotational Semantics for TimeML[*]

Graham Katz

Stanford University

Abstract. The XML-based markup language TimeML encodes temporal and event-time information for use in automatic text processing. The
TimeML annotation of a text contains information about the temporal
intervals that are mentioned in the text as well as the relationship of
these temporal intervals to the times and events mentioned in the text.
We provide here a formal denotational semantics for TimeML, addressing
problems of operator scope that arise in the context of a "flat" representation language and providing a sketch of an intensional extension to
the main extensional semantics.

1 Introduction

TimeML is an XML-based markup language for encoding temporal and event-
time information for use in automatic text processing. The TimeML annotation
of a text contains information about what times and events are mentioned in a
text, as well as information about the temporal relationships that hold among
these times and events. In essence, TimeML is a simple semantic representation language for natural language texts, limited to representing temporal and
event-based information. TimeML is intended to capture the kind of information
conveyed in a text that one might put on a "time-line"—essentially what happened when. TimeML markup thus provides semantic information which might
well be useful for a wide range of applications in which temporal information
is of crucial interest, such as question answering and text summarization [1].
TimeML has been used to hand-annotate a small corpus of newswire texts, and
this annotated corpus, TIMEBANK [2] is now being used as a standard for evaluating the performance of computational systems for doing automatic temporal
interpretation [3].

To give an example, a TimeML annotation of the sentence *"The plane crashed
into the hillside at 10am yesterday"* (the type of sentence found in TIMEBANK)
is given below:

(1) `The plane`
    `<EVENT class="OCCURRENCE" eid="e1" stem="crash"> crashed </EVENT>`
    `into the hillside`
    `<SIGNAL sid="s1"> at </SIGNAL>`

---

[*] This research reported here was carried out in part at the Institute for Cognitive Science at the University of Osnabrück. Thanks to Kai-Uwe Kühnberger, Jens Michaelis
and Peter Bosch, as well as to Cleo Condoravdi at PARC, for helpful discussion

F. Schilder et al. (Eds.): Reasoning about Time and Events, LNAI 4795, pp. 88–106, 2007.
© Springer-Verlag Berlin Heidelberg 2007

```
<TIMEX3 tid="t2" type="TIME" value="2003-11-24-T10:00">
10am  yesterday </TIMEX3>
<MAKEINSTANCE eventID="e1" eventInstanceID="ei1"
 tense="PAST" aspect="NONE"/>
<TLINK lid="l1" eventInstanceID="ei1" signalID="s1"relatedToTime="t2"
relType="IS_INCLUDED"/>
```

The annotation specifies both the actual time that the temporal expression 10am yesterday is intended to refer to, as well as the relationship between this time and the time of the crash. This is information that any English-speaker reading this text in context would derive from it quite naturally. TimeML has been designed to be expressive enough to encode most of the temporal information present in natural language texts [4], and the annotation guidelines for TimeML [5] specify quite specifically how annotators should translate their semantic intuitions into TimeML markup. Although there is obviously an intuitive semantic interpretation associated with these annotation guidelines, the TimeML specification does not explicitly provide a formal, model-theoretic semantics for the TimeML language.[1] In this paper we will take up this task, specifying a model-theoretic semantic interpretation for the TimeML markup language.[2]

The semantics we will propose here is based essentially on the kind of semantic interpretation familiar from the Discourse Representation Theory literature [9,10]. We first provide a straightforward first-order interpretation, treating TimeML markup as a logical representation language, in many ways similar to the Discourse Representations of DRT. This will be seen to have a number of problems related to the scoping of operators such as negation. It should be pointed out that the TimeML language is primarily designed not as a representation language, but as a markup language, meaning that issues concerning ease of use by annotators, correspondence to standards for annotation languages and information processing considerations have played a central role in the design of TimeML. Our task, then, is to fill in *post hoc* the implicit semantics of the intended interpretation of the TimeML language. A number of well-articulated semantic representational languages for representing temporal information have been proposed (for example [11,12]), but these have generally been applied to a limited range of cases, and their use for the analysis of even very short texts has proved difficult [13].

2   TimeML Markup

A TimeML annotated document can be viewed as a list of TimeML tags. There are a number of types of tags. The four main tags relevant for semantic interpretation are the following: <EVENT>, <TIMEX3>, <MAKEINSTANCE> and <TLINK>. These are all illustrated in the above example (other tags, such as the <SLINK> and <ALINK> tags will be discussed later; the <SIGNAL> tag carries no semantic

---

[1] Hobbs & Pustejovsky have made programatic suggestions in this direction [6].

[2] This task is, in many ways, related to the task of specifying a formal semantics for timeline-style diagrams, a problem that has recieved some attention [7,8].

import and will be ignored). Tags carry different types of semantic information. <EVENT> tags mark expressions in the text that refer to event types, <TIMEX3> tags mark relevant time referring expressions, and provide these with a decontextualized value, <MAKEINSTANCE> relates event types to particular event instances <TLINK> tags specify temporal relations among event instances and between events instances and times.

As is typical of markup languages, some tags are used to provide information about bits of text. The <EVENT> tags are associated with event-denoting expressions (such as verbs and event nominals), while the <TIMEX3> tags are associated with temporal expressions such as temporal adverbials. The <MAKEINSTANCE> and <TLINK> tags, however, are "non-consuming" tags which are not associated with any textual material. All tags contain attribute specifications (such as the value specification on the <TIMEX3> tag above). These attributes specify the semantic information encoded in the TimeML markup. In fact, all information relevant to semantic interpretation is contained in the tag labels. As we shall see, this allows us to state the truth conditions for TimeML documents entirely in terms of conditions on the tag labels. The tagged text is of no semantic import.[3]

One of the main features of the TimeML tags is the use of time and event identifier variables in the attribute specifications which can be used as placeholders for underspecified semantic values. These play the role of variable-like elements which can participate in multiple predications. The identifier e1 in example (1) above is an identifier for the crashing event types, the identifier t1 an identifier associated with the time 10am on November 24, 2003, and the identifier ei1 is an identifier associated with the particular crashing event described. Making use of these identifiers is what allows us to relate events with times (via the <TLINK> tag) even in cases in which the time value is unknown. It also allows us to specify the temporal relationships that hold among events without specifying when the events occur.

Note that we have two kinds of identifiers associated with events: event-type identifiers and event-instance identifiers. Distinguishing event instances from event types allows us to make sense of the kind of claim made in a sentence such as *Peter didn't leave*, in which the non-existence of particular events of a general type is what is conveyed, or of sentences such as *Peter played tennis on Monday and on Thursday*, in which a single expression is associated with two actual event instances. This will be discussed in detail below.

3 Semantics for TimeML

3.1 Events and Times

In the tradition of [14], we take events to be concrete individuals located at particular times and in particular places, with particular causal characteristics

---

[3] Version 2.1 of TimeML has included the stem feature on the <EVENT> tag type, which can and will be used to specify the semantic type of the event.

and standing in mereological relations to one another [15,16]—a particular house-building might, for example include a building of a fireplace as a part. Events are sorted into event types, such as crashing events or speaking events. For simplicity we will assume that event types are modeled as sets of event instances. The semantic content of an event-type predicate is to specify the appropriate set. We assume that the lexical item whose stem form is the value of the stem attribute of an <EVENT> tag specifies this. (Of course a more articulated ontology of event types, such as that implicit in the WordNet hierarchy [17] or more explicitly pursued as part of the Semantic Web initiative [18] would be more appropriate.) Here we will simply take each verb root to be associated with an appropriate set of concrete events.

We take <TIMEX3> tags to be associated also with concrete temporal individuals. Following [19] the TimeML annotation essentially adopts the ISO 8601 standard specification for times which defines the modern clock/calendar system in terms of a standardized notation. This is the YYYY-MM-DD-THH:MM notation, which we have already seen as the value on the <TIMEX3> tag above. Formally, we will model time as a set of intervals. The ISO standard specifies two varieties of temporal objects, periods—which we will take to be sets of temporal intervals of equal duration—and non-periods—which we will simply take to be temporal intervals. The association of a particular ISO notation, for example 2003-11-24-T10:00 with a particular interval is defined by the notation itself and is operationalized in terms of the temporal arithmetic defined on it. The fact, for example, that 2003-11-24-T10:00 plus PT27:00 is 2003-11-25-T13:00 is part of this specification. We will not be concerned further with the issue of temporal ontology here, but see [20].

3.2 Models for TimeML Texts

The syntax of the TimeML language is specified formally elsewhere [21], and we will assume this throughout. We will interpret TimeML texts with respect to a class of model structures $\langle E, \preceq, I, <, \subseteq, \tau, Val \rangle$ containing a domain of concrete events and a domain of temporal intervals, where:

E is the set of events,
$\preceq$ is the part of relation on events,
I the set of time intervals,
$<$ is the ordering relation on time intervals,
$\subseteq$ is the inclusion relation on time intervals,
τ is the run-time function from E to I,
Val is the valuation function.

These models must satisfy a number of axioms which capture the intuitive relationship between events, times and ordering. For example, we assume that the ordering relation and inclusion relations are transitive, and that ordering and inclusion are related in the natural way [22,23]. Furthermore, if one event is a part of another, the runtime of the former is included in that of the later.

$$\forall x, y, z \in I[x < y \land y < z \to x < z]$$
$$\forall x, y, z \in I[x \subseteq y \land y \subseteq z \to x \subseteq z]$$
$$\forall w, x, y, z \in I[x < y \land z \subseteq x \land w \subseteq y \to z < w]$$
$$\forall w, x, y, z \in I[x < y \land y < z \land x \subseteq w \land z \subseteq w \to y \subseteq w]$$
$$\forall x, y \in E[x \preceq y \to \tau(x) \subseteq \tau(y)]$$

These axioms specify a fairly simple first order model, whose domain is structured in an intuitively natural way.

TimeML, then, might be thought of as a simple first-order language. The terms are the identifiers, and each tag specifies a property or relation that holds of these terms. The intended models for TimeML are models in which the valuation function Val assigns appropriate denotations for the constants of the language, these being the TimeML tag attributes and their values. For example, the terms which fill the value of the of the `value` attribute of a `<TIMEX3>` tag are character strings, which correspond to the temporal specification of the ISO 8601 standard for time specification. An appropriate valuation function for TimeML models is one that assigns particular intervals to these strings, consistent with the intent specified by the ISO standard itself. Similarly for the constants associated with event predicates. As mentioned above, we assume that the `root` attribute of the `<EVENT>` tag takes as its value symbols that are associated with natural categories of events in an articulated ontology.

We specify this intended interpretation as function as follows. A model **M** is appropriate for interpreting TimeML texts iff:

If α is an ISO-8601 term that doesn't start with P then Val(α) = the set of intervals I' ($\subseteq$ I) which ISO notation specifies for α

If α is an ISO-8601 term that start with P then Val(α) = the set of intervals I' ($\subseteq$ I) such that each i $\in$ I' is of equal length and that is the length determined by the ISO notation

If α is an event predicate then Val(α) = the set of events E' ($\subseteq$ E) such that each e $\in$ E' is an event of the type naturally associated with α.

In short we assume that our temporal value specifications refer to the times that standard ISO-8601 temporal ontology says they should refer to and that the event predicates specifications pick out sets of events of the appropriate type.

To illustrate we will assume that Val(2003-11-24) is the (singleton) set containing the interval of time that is one day long that starts at midnight on the 23rd of November, 2003 and ends a day later. This element will be a member of the set Val(P1D), which will be the set of all day-long intervals. The temporal `value` specification conventions allow for expressions such as XXXX-XX-24 which would have as interpretation the set of all day-long intervals which are the 24th day of some month. Using this as our basic foundation, the central task now is to specify a recurse definition of satisfaction in a model for a complete Time ML text.

3.3   Satisfaction of TimeML Text in Model

Intuitively, a TimeML text is satisfied by an appropriate model if we can find a set of times and event instances which satisfy all the conditions implicit in the tags. For example, in our example TimeML text (1) is satisfied in a model if there is a crashing event whose run time was the interval of time associated with 2003-11-24-T10:00. Our task is to provide a general definition of satisfaction that captures this intuition.

The first thing we have to specify is the interpretation of the time and event identifiers. Recall that these identifiers, such as e1 and t1 in the example, stand for times and events. We will treat these identifiers as variables, and specify their semantics via an embedding function, which will specify for every identifier in the text, what it refers to. For convenience let us define the following functions from TimeML texts to sets of identifiers and tags:

Let T be a TimeML text,
$\quad$ $\text{Dom}_e(T)$ = the set of event ids in T
$\quad$ $\text{Dom}_t(T)$ = the set of time ids in T
$\quad$ $\text{Dom}_{ei}(T)$ = the set of event instance ids in T
$\quad$ $\text{Ident}(T) = \text{Dom}_e(T) \cup \text{Dom}_t(T) \cup \text{Dom}_{ei}(T)$
$\quad$ $\text{Tag}(T)$ = the set of all tags in T

An embedding function is, of course, a function from the set of identifiers to events, times and event types. We can specify the embedding function f as consisting of the union of the following:

$\quad$ $f_e \colon \text{Dom}_e(T) \to \text{Pow}(E),$
$\quad$ $f_{ei} \colon \text{Dom}_{ei}(T) \to E,$
$\quad$ $f_t \colon \text{Dom}_t(T) \to \text{Pow}(I),$

where f_{ei} is one-to-one, meaning that distinct event instance identifiers are mapped to distinct events. All time identifiers are interpreted as sets (perhaps singleton sets) of temporal intervals. We take f: $\text{Ident}(T) \to \text{Pow}(E) \cup E \cup \text{Pow}(I)$ to specify these extensions.

We can now say that a text is satisfied by a model if we can find an embedding function f which satisfies each tag in the text.

A text T is satisfied by a model **M** iff there is a function f (assigning denotations to identifiers) such that for all tags t $\in$ Tag(T), t is satisfied by f in **M**.

We need, of course, to specify what it means for a tag to be satisfied by an assignment function in a model. We specify this by enumeration.

We first enumerate tag satisfaction conditions for the <EVENT> and <TIMEX3> tags. Here it is the identifiers that play the central role in the interpretation. Essentially, an <EVENT> or <TIMEX3> tag is satisfied by an embedding function f if the identifier is assigned the appropriate interpretation by f, in the case of an event identifier this will be as a set of events denoted by the root attribute, in the case of a temporal identifier, this will the set of times denoted by the value attribute.

A tag **t** is satisfied by an embedding function f in **M** iff if **t** has the form
    <EVENT eid=α class=β root=γ > then f(α) = Val(γ),
    <TIMEX3 tid=α value=γ > then f(α) = Val(γ), ...

We now enumerate the tag-satisfaction conditions for the <MAKEINSTANCE> tag. This tag specifies the relationship between an event type and an event token.[4]

if **t** has the form
    <MAKEINSTANCE eiid=α eid=β polarity="POS" modality= "">
    then f(α) $\in$ f(β),
    <MAKEINSTANCE eiid=α eid=β polarity="NEG" modality= "">
    then f(α) $\notin$ f(β), ...

A positive <MAKEINSTANCE> tag is satisfied if the event instance identifier is interpreted as an event of the type which the event identifier is interpreted, while a negative <MAKEINSTANCE> tag is satisfied if this is not the case. We will address the modality attribute below, as interpreting <MAKEINSTANCE> tags with non-null modality requires us to enrich our models significantly.

Finally, we consider the non-consuming <TLINK> tags, which relates times and events temporally. There are 28 types of <TLINK>s (14 relation types relating an event instances to another event instance, and 14 relation types relating an event instance to a time). I will only give satisfaction conditions for three exemplary <TLINK> types here. It should be clear what the clauses for the remaining <TLINK>-types are. Let us consider first links relating event instances. These are straightforward—the temporal relation specified by the <TLINK>s relType attribute indicates the relationship that holds between the run times of the events related:

if **t** has the form:
    <TLINK eventInstanceID=α relatedtoEventInstance=β
    relType= "IS_INCLUDED">
    then $\tau(f(\alpha)) \subseteq \tau(f(\beta))$
    <TLINK eventInstanceID=α relatedtoEventInstance=β
    relType= "BEFORE">
    then $\tau(f(\alpha)) < \tau(f(\beta))$
    <TLINK eventInstanceID=α relatedtoEventInstance=β
    relType= "DURING">
    then $\tau(f(\alpha)) = \tau(f(\beta))$

Next we turn to the <TLINK> between events instances and time. This is made only slightly more complex by the fact that temporal identifiers are are interpreted as sets of intervals. Relations here will be specified as holding between at least one element of the set.

---

[4] We are only considering the case in which the cardinality is unspecified (i.e. corresponds to 1). The semantics for other values of the tag feature cardinality is highly problematic.

```
<TLINK eventInstanceID=α relatedtoTime=β
relType= "IS_INCLUDED">
```
then $\exists\ I \in f(\beta)$ such that $\tau(f(\alpha)) \subseteq I$,
```
<TLINK eventInstanceID=α relatedtoTime=β
relType= "BEFORE">
```
then $\exists\ I \in f(\beta)$ such that $\tau(f(\alpha)) < I$,
```
<TLINK eventInstanceID=α relatedtoTime=β
relType= "DURING">
```
then $\exists\ I \in f(\beta)$ such that $\tau(f(\alpha)) = I$,

<TLINK> tags are in general satisfied by an embedding function if the identifiers which are related are mapped to events or times which stand in the appropriate temporal relation.

Let us illustrate the system as we have developed it so far by applying it to the interpretation of the following very brief TimeML text.

(2) John
```
<EVENT eid="e1"    class="OCCURRENCE"   pred="TEACH">
taught
</EVENT>
<TIMEX3 tid="t1" type="DURATION" value="P20M">
20 minutes
</TIMEX3>
<SIGNAL sid="s1">
on
</SIGNAL>
<TIMEX3 tid="t2" type="DATE" value="XXXX-WXX-1">
Monday
</TIMEX3>
<MAKEINSTANCE eventID="e1" eventInstanceID="ei1" " negation="FALSE">
<TLINK eventInstanceID="ei1" signalID="s1" relatedToTime="t2"
relType="IS_INCLUDED"/>
<TLINK eventInstanceID="ei1" relatedToTime="t1" relType="DURING"/>
```

The first thing we need to do is to specify, for this text, what domain of identifiers of each type is. For such a short text this is fairly straightforward:

$$\mathrm{Dom}_e(T_2) = \{e1\}$$
$$\mathrm{Dom}_e i(T_2) = \{ei1\}$$
$$\mathrm{Dom}_t(T_2) = \{t1,t2\}$$

Then we can determine whether or not the annotation is satisfied in a model **M**. It will be satisfied if we can find an embedding function f such that:

$f(e1) = \mathrm{Val}(\texttt{teach})$ *the set of teaching events*
$f(t2) = \mathrm{Val}(\texttt{XXXX-WXX-1})$ *the set of Mondays*
$f(t1) = \mathrm{Val}(\texttt{P20M})$ *the set of 20 minute intervals*
$f(ei1) \in f(e1)$
there is an $i \in f(t2)$ such that $\tau(f(ei1)) \subseteq i$
there is an $i \in f(t1)$ such that $\tau(f(ei1)) = f(t1)$

This seems to give the correct truth conditions, which are essentially those of the following first-order formula.

$$\exists\ e,t,t'\ [\text{teaching}(e) \wedge \tau(e) = t \wedge t \subseteq t' \wedge \text{monday}(t') \wedge 20\text{min}(t)]$$

This is, of course, the intended interpretation. Note that the implicit existential associated with the embedding function is what gives rise to the existential quantification over times. In contrast, the explicit existential associated with the <MAKEINSTANCE> tag is what gives rise to the existential quantification over events.

It may already be clear that there are significant problems for this analysis, and we will come to these shortly. But before addressing these, let us turn to the tag types <ALINK> and <SLINK>.

3.4 Secondary Link Tags

In addition to the temporal <TLINK> tags, TimeML has two other kinds of non-consuming "link" tags which relate events: aspectual <ALINK> tags and modal <SLINK> tags. The <ALINK> tags are used to relate events to their aspectual parts and the <SLINK> tags are used to relate events to other modally subordinated events. An <ALINK> is illustrated in the annotated sentence (3) below and a <SLINK> in (4).

```
(3) They
    <EVENT eid="e1" class="ASPECTUAL" root="begin">
    began
    </EVENT>
    <MAKEINSTANCE eiid="ei1" eventID="e1" tense="PAST"
    aspect="NONE" polarity="POS"/>
    <EVENT eid="e1" class="OCCURRENCE" root="withdraw">
    withdrawing
    </EVENT>
    <TIMEX3 tid="t1" type="DATE" value="XXXX-WXX-1">
    Monday
    </TIMEX3>
    <MAKEINSTANCE eiid="ei2" eventID="e2" tense="nil"
    aspect="PROG" polarity="POS"/>
    <ALINK eventInstanceID="ei1" signalID="s1"
    relatedToEvent="ei2" relType="INITIATE"/>
    <TLINK eventInstanceID="ei1" relatedToTime="t1"
    relType="IS-INCLUDED"/>

(4) Bill
    <EVENT eid="e1" class="I_STATE" root="want">
    wants
    </EVENT>
    <MAKEINSTANCE eiid="ei1" eventID="e1" polarity="POS"/>
    to
    <EVENT eid="e2" class="OCCURRENCE" root="teach">
```

```
teach
</EVENT>
<MAKEINSTANCE eiid="ei2" eventID="e2" polarity="POS"/>
<TIMEX3 tid="t1" type="DATE" value="XXXX-WXX-1">
Monday
</TIMEX3>
<TLINK eventInstanceID="ei2"  relatedToTime="t1"
relType="IS_INCLUDED"/>
<SLINK eventInstanceID="ei1" subordinatedEventInstance="ei2"
relType="MODAL"/>
```

Interpreting <SLINK> tags also requires us to consider more enriched models, and we will set them aside for the moment. <ALINK>s are less problematic. The <ALINK> tag above, for example, is used to relate the beginning of the withdrawal and the withdrawal itself, a relation that is fairly straightforward to express in terms of the event-part relation $\preceq$ and the temporal relation $<$.

There are five types of <ALINK> relations: INITIATES, CULMINATES, TERMINATES, CONTINUES and REINITIATES. We can specify the tag satisfaction conditions of these in a straightforward manner, making use of the functions **beg** and **end** which specify the first and last point of a temporal interval (REINITIATES is set aside, as it has primarily modal content).

A tag **t** is satisfied by the embedding function f in M iff if **t** has the form:

> <ALINK eventInstanceID=α relatedtoEventInstance=β
> relType="INITIATES">
> then f(α)) $\preceq$ f(β) and $\mathbf{beg}(\tau(f(\alpha))) = \mathbf{beg}(\tau(f(\beta)))$
> <ALINK eventInstanceID=α relatedtoEventInstance=β
> relType="CULMINATES"> or
> <ALINK eventInstanceID=α relatedtoEventInstance=β
> relType="TERMINATES">
> then f(α)) $\preceq$ f(β) and $\mathbf{end}(\tau(f(\alpha))) = \mathbf{end}(\tau(f(\beta)))$
> <ALINK eventInstanceID=α relatedtoEventInstance=β
> relType="CONTINUES">
> then f(α)) $\preceq$ f(β) and $\tau(f(\alpha)) \subseteq \tau(f(\beta))$

The annotated text above is satisfied in a model M if we can find an embedding f such that:

f(e1) = Val(begin)	*the set of beginning events*
f(e2) = Val(withdraw)	*the set of withdrawal events*
f(t1) = Val(XXXX-WXX-1)	*the set of Mondays*
f(ei1) $\in$ f(e1)	
f(ei2) $\in$ f(e2)	
τ(f(ei1)) $\subseteq$ f(t1)	
τ(f(ei1)) $\subseteq$ τ(f(ei2))	
f(ei1) $\sqsubset$ f(ei1)	

Again this seems to be about what is desired. There is, of course, much to say about the detailed semantics of these aspectual verbs, which we are ignoring here [24].

For a wide range of cases, given an appropriate interpretation for the constants, this semantics for this subset of TimeML delivers exactly the intuitive truth conditions for which it was designed, essentially specifying what happened when and what happened before or during what. There are, however, a number of issues that arise when we begin to consider even a slightly wider range of data, and these have to do with cases in which the information conveyed in the text is not simply positive and conjunctive. We consider these in the next section.

4   Problems of Operator Scope

Not all information is positive and conjunctive. The following sentence from TIMEBANK, for example, tell us about the non-existence of certain events:

```
(5) It hasn't
    <EVENT eid="e490" class="OCCURRENCE"> diversified </EVENT>
    beyond steel,  nor has it
    <EVENT eid="e56" class="OCCURRENCE"> linked up </EVENT>
    with a joint venture partner to
    <EVENT eid="e59" class="OCCURRENCE">share</EVENT>
    costs and risks.
```

While "negative" events such as the non-occurrence of the diversification described above are relatively rare they are not non-existent. In TIMEBANK, 295 of the 7940 <MAKEINSTANCE> tags have NEG polarity. To illustrate, let us consider the interpretation of the somewhat simplified example given below, in which we have temporal expressions linked to a single negative event.

```
(6) John didn't
    <EVENT eid="e1" class="OCCURRENCE"  stem="teach">
    teach
    </EVENT>
    <SIGNAL sid="s1">
    on
    </SIGNAL>
    <TIMEX3 tid="t2" type="DATE" value="XXXX-WXX-1">
    Monday
    </TIMEX3>
    <MAKEINSTANCE eventID="e1" eventInstanceID="ei1" polarity="NEG">
    <TLINK eventInstanceID="ei1" signalID="s1" relatedToTime="t2"
    relType="IS-INCLUDED"/>
```

Following the semantic interpretation we gave above, this TimeML text is satisfied in a model M if we can an embedding function f such that:

$$f(\texttt{e1}) = \text{Val}(\texttt{teach}) \qquad \textit{the set of teaching events}$$
$$f(\texttt{t2}) = \text{Val}(\texttt{XXXX-WXX-1}) \qquad \textit{the set of Mondays}$$

$$f(ei1) \not\subseteq f(e1)$$
$$\exists\, I \in f(t2)\ \tau(f(ei1)) \subseteq I$$

In other words, the text is satisfied if there was an event that was not a teaching that was on a Monday. This is clearly incorrect, since a model in which there is a non-teaching on a Monday might also be one in which there is a teaching on a Monday. Again the "translation" to first order formulae might be helpful. What we in fact have is an interpretation equivalent to:

$$\exists\, e,t\ [\neg\text{teaching}(e) \land \tau(e) \subseteq t \land \text{monday}(t)]$$

What we would like, however, is something equivalent to:

$$\neg\exists\, e,t\ [\text{teaching}(e) \land \tau(e) \subseteq t \land \text{monday}(t)]$$

The basic problem is that TimeML provides no mechanism for indicating scopal relations. In cases like that given above negation should take scope both over the temporal relation encoded by the `<TLINK>` tag and over the event type predicate indicated by the `<MAKEINSTANCE>` tag, but as it is the single tags are interpreted conjunctively and negation can only scope over one of these atomic predications, here that associated with the `<MAKEINSTANCE>` tag.

This problem extends to any case in which such scoping mechanisms are needed. For example, the subordinating `<SLINK>` tags also require a scope mechanism, as these links are intended to convey an informational subordinating relationship. In the case we gave above, for example, this would be the relationship between Bill's desires (the "wanting" event) and a potential event of teaching on Monday. Here again there is no actual event of teaching on Monday that is being denoted, and the description for this potential event needs to be assembled out of a number of different TimeML tags.

That sentences whose semantic interpretation crucially involves expressions which are traditionally taken to be scope taking operators should be problematic is not entire surprising: the TimeML annotation is entirely "flat"—TimeML tags don't contain any embedding—and therefore if we interpret them, as we have, as simple first-order conjunctions we are bound to run into trouble. Under its "natural" interpretation— which we have formalized above—TimeML is a very week logical representation language—essentially DRT without embedded DRSs.[5]

4.1 Simulating Scope

We would like to interpret the following set of tags (which would be associated with the sentence *John didn't teach for 20 minutes on Monday*) with the interpretation given informally in first-order notation below.

---

[5] Since TimeML was designed to encode the kinds of information that simple timeline diagrams do, it should not be a surprise if we find it to have the same kinds of weaknesses that diagrams have been shown to have [25,26].

(7) `<EVENT eid= "ei1" root="teach">`
   `<TIMEX3 tid="t1" val="XXXX-WXX-1">`
   `<MAKEINSTANCE eiid="ei1" eventID="e1" polarity="NEG">`
   `<TLINK relatedToTime ="ti1" eventID="ei1" relation="IS_INCLUDED">`

(8) $\neg \exists$ e,t [teaching(e) $\wedge$ τ(e) $\subseteq$ t $\wedge$ monday(t) $\wedge$ 20min(t)]

What we don't have, however, is a way of letting the negation scope over the entire clause. This lack of a "scope domain" is the central representational short-coming of TimeML. In this section we sketch an approach to addressing this problem.

Our approach will be to treat identifiers not as standing for event instances, but rather to treat them as standing for predicates of events. Before giving this revised interpretation of the TimeML language, let us illustrate the idea by way of the second-order formulae below. The essential idea is that each of the conjoined conditions (contributed by the individual tags) will partially specify an event predicate. The temporal relations will go into constructing this event predicate, and the polarity of the sentence will determine if there is or is not an event of which this predicate holds. In order to specify this notion we will use the following defined logical operators: the $\sqsubseteq$ to relate predicates[6] and we will use a maximizing existential $\exists_{max}$[7]. This will allow us to specify the maximal predicate which entails each of the relations specified by tags, and this is the central idea behind the analysis. Each tag will be taken to contribute conjunctive information about an event predicate. The tags will be satisfied if there is (in the positive case) or is not (in the negative case) an event of which the maximal event predicate subsuming each of the tag satisfaction conditions associated with the tag holds.

We illustrate this below:

(9) John taught for 20 minutes on Monday.
   $\exists_{max}$ P [P $\sqsubseteq$ λ e [teaching(e)] $\wedge$ P $\sqsubseteq$ λ e [20min(τ(e))] $\wedge$ P $\sqsubseteq$ λ e [τ(e) $\subseteq$ Monday)]] $\wedge$ $\exists$ e [P(e)]

(10) John didnt teach for 20 minutes on Monday
   $\exists_{max}$ [P $\sqsubseteq$ λ e [teaching(e)] $\wedge$ P $\sqsubseteq$ λ e [20min(τ(e))] $\wedge$ P $\sqsubseteq$ λ e [τ(e) $\subseteq$ Monday)]] $\wedge$ $\neg\exists$ e [P(e)]

In this case, then, we specify a predicate which is true of (all) events which are teachings, which are 20 minutes long and which occur on Monday and specify either that there is such an event (or that there is not.

4.2   Type-Level Satisfaction

In order to implement this idea, we need to provide a new definition of satisfaction. We do this by modifying the satisfaction definition slightly. As above, we take the domain of the embedding functions to be the various sets of identifiers in the text. So, if we let T be a TimeML text, then:

---

[6] $\forall$ P,Q [P $\sqsubseteq$ Q $\leftrightarrow$ $\forall$ x P(x) $\rightarrow$ Q(x).
[7] $\exists_{max}$ P ϕ is true iff $\exists$ P ϕ and $\neg\exists$ P' [P $\sqsubseteq$ P' $\wedge$ ϕ[P'/P]].

$\text{Dom}_e(T) = $ the set of event ids in T
$\text{Dom}_{ei}(T) = $ the set of event instance ids in T
$\text{Dom}_t(T) = $ the set of time ids in T
$\text{Tag}(T) = $ the set of all tags in T

An embedding function f will be taken to be the union of the functions:

$f_e \colon \text{Dom}_e(T) \rightarrow \text{Pow}(E),$
$f_{ei} \colon \text{Dom}_{ei}(T) \rightarrow \text{Pow}(E),$
$f_t \colon \text{Dom}_t(T) \rightarrow \text{Pow}(I)$

Note that in contrast to the previous section, we take f_{ei} to map event instance identifiers to sets of events. We now say that a text T is satisfied by a model M iff there is a maximal embedding function f such that f every tag in T is satisfied by f. An embedding function f is maximal (in the sense here intended) iff $\forall$ x,y $[f(x) = y] \rightarrow \forall$ f' $[f'(x) \subseteq y]$. This means that when f assigns an extension to an element, it always assigns the maximal possible extension.

The satisfaction conditions for <EVENT> tags and <TIMEX3> tags need not be altered. The new satisfaction conditions for <TLINK>, <ALINK>, and <MAKEINSTANCE> tags will be parallel to those we gave above, but will always specify a partial specification of an event predicate. So, for example the <TLINK> relation will specify that identifier associated with the eventInstanceID attribute is mapped to a maximal subset of the set of events which stand in the appropriate relation to a time associated with the relatedToTime identifier. In general, then, a tag **t** is satisfied by f in **M** iff :

if **t** has the form:
 <TLINK eventInstanceID=α relatedtoTime=β
 relType="IS_INCLUDED">
 then there is an $i \in I$ such that $f(\alpha) \sqsubseteq \lambda$ e $[\ \tau(e) \subseteq i]$,
 <TLINK eventInstanceID=α relatedtoTime=β relType="BEFORE">
 then there is an $i \in I$ such that $f(\alpha) \sqsubseteq \lambda$ e $[\ \tau(e) < i]$,
 <TLINK eventID=α relatedtoTime=β relType="DURING">
 then there is an $i \in I$ such that $f(\alpha) \sqsubseteq \lambda$ e $[\ \tau(e) = i]$,

The event-instances are interpreted as sets of events that stand in the appropriate relations. MAKEINSTANCE will now simply be a certain sort of specification:

if **t** has the form:
 <MAKEINSTANCE eventInstanceID=α eventID=β polarity="POS"
 modality="">
 then there is an event e $\in$ f(β) such that e $\in$ f(α)
 <MAKEINSTANCE eventInstanceID=α eventID=β polarity="NEG"
 modality="">
 then there is no event e $\in$ f(β) such that e $\in$ f(α)

In other words, the <MAKEINSTANCE> tag is to be reinterpreted as a relation between two event types, one of these is that related to the eventID (associated with the main lexically specified event-type predicate) and the other related to the eventInstanceID which is the event predicate that is specified by the various <TLINK> tags.

It should be clear that this is explicitly **not** the interpretation that was implicit in the design of TimeML, rather we are here providing a coherent *post-hoc* interpretation of the TimeML language as it stands that addresses the scope problem raised above. These new satisfaction conditions do assure us that the tags in (7) receive the interpretation as in (10) above.

There are a number of expressions—in particular disjunction and modality— which require special treatment along the lines that we have just given for negation because of their scopal properties. In the next section we will address the interpretation of modality features in TimeML. Unlike disjunction, modality is quite common in TIMEBANK and therefore calls for some attention.[8]

4.3 Modality and SLINKs

We will make no effort here to give a serious treatment of natural language modality, but restrict ourselves to sketching how the scope issue associated with modality can be addressed along the lines we have explored above. Modality may be tagged in two places in TimeML, once on the **modality** attribute of the <MAKEINSTANCE> tag and also via the **relType** attribute of the <SLINK> tag. The <SLINK> tag we have illustrated above. Simple modality as marked on <MAKEINSTANCE> is as illustrated in (11).

(11) John can
```
<EVENT eid="e1" class="OCCURRENCE"  stem="teach">
teach
</EVENT>
<SIGNAL sid="s1">
on
</SIGNAL>
<TIMEX3 tid="t2" type="DATE" value="XXXX-WXX-1">
Monday
</TIMEX3>
<MAKEINSTANCE eventID="e1" eventInstanceID="ei1" polarity="NEG" modal
<TLINK eventInstanceID="ei1" signalID="s1" relatedToTime="t2"
relType="IS-INCLUDED"/>
```

In both cases the intent of marking modality is to indicate that the event being specified as modal is not an actual event, but merely a potential event. We will adopt the standard semantic possible worlds treatment of modality [28,29]. Actual events will be those that exist in the actual world, and non-actual those that exist in other possible worlds.

Let the model structures be extended to contain a set $\mathbf{W}$ of possible worlds (with the distinguished world w_0) and the relation $\mathbf{Acc}$ among worlds. The

---

[8] For a different approach see [27].

valuation function Val will provide each `<TIMEX3>` `value` and each `<EVENT>` `root` with an intensional interpretation—a function from possible worlds to sets of times on the one hand and from possible worlds to event predicates on the other. Embedding functions will now map identifiers to intensions (i.e. functions from worlds to event predicates). An embedding function f will be taken to be the union of the functions:

f_e: $\text{Dom}_e(T) \to W \to \text{Pow}(E)$,
f_{ei}: $\text{Dom}_{ei}(T) \to W \to \text{Pow}(E)$,
f_t: $\text{Dom}_t(T) \to W \to \text{Pow}(I)$,

We will say that a text T is satisfied in a model M at a world w iff all tags in the text are satisfied by M in w by a maximal embedding[9] f. A tag t is satisfied by f in **M** at w iff :

if **t** has the form:
    `<TLINK eventInstanceID=`α` relatedtoTime=`β
    `relType="IS_INCLUDED">`
    then there is an $i \in I$ such that $f(\alpha) \sqsubseteq \lambda w \lambda e [\tau(e) \subseteq i]$,
    `<TLINK eventInstanceID=`α` relatedtoTime=`β` relType="BEFORE">`
    then there is an $i \in I$ such that $f(\alpha) \sqsubseteq \lambda w \lambda w \lambda e [\tau(e) < i]$,
    `<TLINK eventID =`α` relatedtoTime =`β` relType= "DURING">`
    then there is an $i \in I$ such that $f(\alpha) \sqsubseteq \lambda w \lambda e [\tau(e) = i]$,

The event-instances are interpreted as sets of events that stand in the appropriate relations. MAKEINSTANCE will now simply be a certain sort of specification of the relation of the the actuality of the event described. In the non-modal case the event described will be be taken to exist in the world of evaluation, while in the modal case it will be taken to hold in another world. So, a tag **t** is satisfied by f in **M** at **w** iff:

if **t** has the form:
    `<MAKEINSTANCE eventInstanceID=`α` eventID=`β
    `polarity="POS" modality= "">`
    then there is an event $e \in f(\beta)(w)$ such that $e \in f(\alpha)(w)$,
    `<MAKEINSTANCE eventInstanceID=`α` eventID=`β
    `polarity="NEG" modality= "">`
    then there is no event $e \in f(\beta)(w)$ such that $e \in f(\alpha)(w)$,
    `<MAKEINSTANCE eventInstanceID=`α` eventID=`β
    `polarity="POS" modality="CAN">`
    then there is a world $w' \in W$ such that $Acc(w,w')$ and there is an event $e \in f(\beta)(w')$ such that $e \in f(\alpha)(w')$,
    `<MAKEINSTANCE eventInstanceID=`α` eventID=`β
    `polarity="NEG" modality="CAN">` then there is a world $w' \in W$ such that $Acc(w,w')$ and there is no an event $e \in f(\beta)(w')$ such that $e \in f(\alpha)(w')$, ...

_____

[9] A maximal embedding is now interpreted as maximal on worlds and on events.

We can extend this analysis to the treatment of `<SLINK>`s by allowing the link to specify the modal relation. We make use of the notion of an e alternatives to a world—these being the worlds which are accessible from w which are compatible with the agent of e's attitude in e. In other words these will be the worlds compatible with Bill's desires, in the case in which e is the event of Bill's wanting expressed in (4). We can now extend this modal semantics to the treatment of `<SLINK>`s. For these the tag is satisfied by f in **M** at **w** iff:

> if **t** has the form:
> `<SLINK eventInstanceID=`α` subordinatedEventInstance=`β
> `relType="MODAL"/>`
> then every world $w' \in W$ which is an alternative to w compatible with $f(\alpha)$ there is an event $e \in f(\beta)(w')$

Here it is the `<SLINK>` itself which indicates that the modal relation to the subordinate event is lexically specified by the subordinating attitude event. It will be the case, then, that (4) is satisfied in a world w_0 in a modal iff every world which is a desire alternative of Bills in w_0 is a world in which there is an event of teaching which is on a Monday. This, of course, is the classical analysis.

What is crucial is how the scope of the modal expression, be it that on the `<MAKEINSTANCE>` tag or that associated with the `<SLINK>` tag, is simulated. This is done by making use of the subsumption of intensions and by requiring that the satisfying embedding be maximal. Intuitively in the treatment of (4) we are relating Bill's desires to subordinate teaching events that are on Monday by "building" a maximal intentional event predicate which subsumes teaching events and monday events and relating this to Bill's desire event.

5 Conclusion

As we have seen, the formal semantic interpretation of the TimeML markup language is not nearly as straightforward as one might have expected from the intuitive characterization implicit in [4]. The characterization of simple positive time and event information is relatively straightforward. We take the event identifiers and the time identifiers introduced by `<EVENT>` or `<TIMEX3>` tags to be interpreted much like variables or discourse referents, which receive more or less specification as to their actual value from the attributes on this tag as well as from the attributes of other TimeML tags in the text. More problematic are natural language expressions such as negation, modality, indirect speech and expressions of propositional attitude, which have traditionally been analyzed as scope taking operators. These present challenges for the interpretation of a language which contains no syntactic reflexes of embedding.

The analysis presented here, in which the semantic effects of embedding structures is reflected the second order characterization of event predicates, has much in common with recent work in Minimal Recursion Semantics [30]. The semantic scope of operators is only simulated and not expressed directly in the representation language, as it is in DRT, for example. Whether this type treatment can be successfully applied to the wide range of data for which TimeML has been designed remains to be seen. It should be clear that while TimeML appears to provide the tools for representing a wide range of temporal and event-based information expressed in natural language, giving this annotation language a well-defined semantics illustrates, once again, how difficult the task of doing natural language semantics truly is.

References

1. Pustejovsky, J., Saurí, R., Castaño, J., Radev, D., Gaizauskas, R., Setzer, A., Sundheim, B., Katz, G.: Representing temporal and event knowledge for qa systems. In: Maybury, M.T. (ed.) New Directions in Question Answering, MIT Press, Cambridge (2004)
2. Pustejovsky, J., Hanks, P., Saurí, R., See, A., Gaizauskas, R., Setzer, A., Radev, D., Sundheim, B., Day, D., Ferro, L., Lazo, M.: The Timebank corpus. In: Proceedings of Corpus Linguistics 2003, Lancaster, pp. 647–656 (2003)
3. Verhagen, M., Gaizauskas, R., Schilder, F., Hepple, M., Katz, G., Pustejovsk, J.: Semeval-2007 task 15: Tempeval temporal relation identification. In: ACL SemEval Workshop (2007)
4. Pustejovsky, J., Ingria, R., Castaño, J., Saurí, R., Littman, J., Gaizauskas, R., Setzer, A., Katz, G., Mani, I.: The specification language timeml. In: Mani, I., Pustejovsky, J., Gaizauskas, R. (eds.) The Language of Time: A Reader, Oxford, pp. 545–557 (2005)
5. Saurí, R., Littman, J., Knippen, B., Gaizauskas, R., Setzer, A., Pustejovsky, J.: Timeml annotation guidelines version 1.2.1 (2006)
6. Hobbes, J.R., Pustejovsky, J.: Annotating and reasoning about time and events. American Association for Artificial Intelligence (2003)
7. Dillon, L.K., Kutty, G., Melliar-Smith, P.M., Moser, L.E., Ramakrishna, Y.S.: Visual specifications for temporal reasoning. Journal of Visual Languages and Computing 5(1), 61–81 (1994)
8. Smith, M.H., Holzmann, G.J., Etessami, K.: Events and constraints: A graphical editor for capturing logic requirements of programs. In: RE 2001. Proceedings of the Fifth IEEE International Symposium on Requirements Engineering (RE 2001), p. 14. IEEE Computer Society, Washington (2001)
9. Kamp, H.: A theory of truth and semantic representation. In: Groenendijk, J., Janssen, T., Stokhof, M. (eds.) Formal Methods in the Study of Language, Mathematical Centre, Amsterdam (1981)
10. Kamp, H., Reyle, U.: From Discourse to Logic. Kluwer Academic Publishers, Dordrecht (1993)
11. Hwang, C., Schubert, L.K.: Interpreting tense, aspect, and time adverbials: a compositional, unified approach. In: Gabbay, D., Ohlbach, H. (eds.) Proc. of the 1st Int. Conf. on Temporal Logic, pp. 238–264 (1994)
12. ter Meulen, A.G.B.: Representing time in natural language: the dynamic interpretation of tense and aspect. MIT Press, Cambridge (1995)
13. Reyle, U., Roßdeutscher, A.: Temporal underspecification in discourse. In: Rohrer, A.R.C., Kamp, H. (eds.) Linguistic Form and its Computation, CSLI Publications, Stanford, CA (2001)
14. Davidson, D.: The logical form of action sentences. In: Rescher, N. (ed.) The Logic of Decision and Action, Pittsburgh Press, Pittsburgh (1967)

15. Link, G.: Algebraic semantics for event structures. In: Stokhof, M., Veltman, F. (eds.) Proceedings of the Sixth Amsterdam Colloquium, University of Amsterdam, Institute for Language Logic and Information, pp. 243–262 (1987)
16. Krifka, M.: Nominal reference, temporal constitution and quantification in event semantics. In: Bartsch, R., van Benthem, J., van Emde Boas, P. (eds.) Semantics and Contextual Expression, Foris Publications, Dordrecht (1989)
17. Miller, G.A.: Wordnet: A lexical database for english. Commun. ACM 38(11), 39–41 (1995)
18. Shadbolt, N., Lee, T.B., Hall, W.: The semantic web revisited. IEEE Intelligent Systems 21(3), 96–101 (2006)
19. Ferro, L., Mani, I., Sundheim, B., Wilson, G.: Tides temporal annotation guidelines draft - version 1.02. Mitre technical report mtr mtr 01w000004, The Mitre Corporation, McLean, Virginia (2001)
20. Hobbes, J.R., Pan, F.: An ontology of time for the semantic web. ACM TRansactions on Asian Languages Information Processing 3, 66–85 (2004)
21. Group, T.W.: Timeml 1.2.1 a formal specification language for events and temporal expressions
22. Van Bentham, J.: The Logic of Time. Reidel, Dordrecht (1983)
23. Landman, F.: Structures for Semantics. Kluwer, Dordrecht (1991)
24. Freed, A.F.: The Semantics of English Aspectual Complementation. Reidel, London (1979)
25. Lemmon, O.: Comparing the efficacy of visual languages. In: Barker-Plummer, D., Beaver, D.I., van Benthem, J., di Luzio, P.S. (eds.) Words, Proofs and Diagrams, CSLI Publications (2002)
26. Lemmon, O., Pratt, I.: On the insufficiency of linear diagrams for syllogisms. Notre Dame Journal of Formal Logic 39, 573–580 (1998)
27. Saurí, R., Verhagen, M.: Temporal information in intensional contexts. In: Bunt, H., Geertzen, J., Thijse, E. (eds.) IWCS-6. Sixth International Workshop on Computational Semantics, pp. 404–406 (2005)
28. Kripke, S.: A completeness theorem in modal logic. Journal of Symbolic Logic 24, 1–14 (1959)
29. Hintikka, J.: Semantics for propositional attitudes. In: Davis, J.W. (ed.) Philsophical Logic, pp. 21–45. Reidel, Dordrecht (1969)
30. Copestake, A., Flickinger, D., Pollard, C., Sag, I.A.: Minimal recursion semantics: an introduction. Research on Language and Computation 3(4), 281–332 (2006)

Arguments in TimeML: Events and Entities

James Pustejovsky, Jessica Littman, and Roser Saurí

Computer Science Department, Brandeis University
415 South St., Waltham, MA 02454 USA
{jamesp,jlittman,roser}@cs.brandeis.edu

Abstract. TimeML is a specification language for the annotation of
events and temporal expressions in natural language text. In addition,
the language introduces three relational tags linking temporal objects
and events to one another. These links impose both aspectual and temporal ordering over time objects, as well as mark up subordination contexts
introduced by modality, evidentiality, and factivity. Given the richness of
this specification, the TimeML working group decided not to include the
arguments of events within the language specification itself. Full reasoning and inference over natural language texts clearly requires knowledge
of events along with their participants. In this paper, we define the appropriate role of argumenthood within event markup and propose that
TimeML should make a basic distinction between arguments that are
events and those that are entities. We first review how TimeML treats
event arguments in subordinating and aspectual contexts, creating event-
event relations between predicate and argument. As it turns out, these
constructions cover a large number of the argument types selected for by
event predicates. We suggest that TimeML be enriched slightly to include
causal predicates, such as *lead to*, since these also involve event-event relations. As such, causal relationships will be a relation type for the new
Discourse Link that will also encode other discourse relations such as
elaboration. We propose that all other verbal arguments be ignored by
the specification, and any predicate-argument binding of participants to
an event should be performed by independent means. In fact, except for
the event-denoting arguments handled by the extension to TimeML proposed here, almost full temporal ordering of the events in a text can be
computed without argument identification.

1   Introduction

The question to be addressed in this paper is not *whether* arguments should be
included in the specification language of TimeML, but *which* arguments should
be and *how* they should best be represented. We review the treatment of complex
complementation in TimeML, whereby a proposition-denoting or event-denoting
expression is linked to the predicate (event) introducing it by an explicit relational tag, the SLINK. This effectively binds these complements as arguments
to their governing events. In fact, currently, any event-denoting expression ap-
pearing as an argument to a predicate, broadly speaking, is annotated explicitly

F. Schilder et al. (Eds.): Reasoning about Time and Events, LNAI 4795, pp. 107–126, 2007.
© Springer-Verlag Berlin Heidelberg 2007

in a link relation. In this paper, we wish to make the strategy explicit by which an argument to an event is annotated. We suggest that predicates selecting for situation, proposition, or event types should be part of the explicit annotation of an event. As a result, this requires expanding the specification to include causal predicates and other discourse markers such as *lead to* and *induce*. Finally, we suggest the simplest way to incorporate entity-denoting arguments in the specification. While the specification language allows for bindings to entities, it does not require annotation of the entities for well-formed markup.

2   Overview of Current TimeML Specification

The TimeML specification language provides a standard for capturing all temporal information in a natural language text. This includes temporal expressions, events, and the relationships they share. To achieve such an annotation, TimeML uses four main tag types that fall into two categories, those that consume text and those that do not. TIMEX3, SIGNAL, and EVENT fall into the former group. The LINK tag type is non-consuming as it relates temporal objects to one another. In addition, the specification allows for non-consuming TIMEX3 tags to hold information on implicit temporal expressions, and non-consuming EVENT tags if more than one instance of an event is needed. In the subsections that follow, we briefly describe each of these tags.

2.1   Temporal Expressions

TimeML expands on earlier attempts to annotate temporal expressions ([1], [2]), with the introduction of the TIMEX3 tag. Specifically, TIMEX3 adds functionality to the TIMEX2 standard [3].

Temporal expressions in TimeML fall into four categories: DATEs, TIMEs, DURATIONs, and SETs. A DATE is any calendar expression such as *July 3* or *February, 2005*. The annotation of such examples includes a value attribute that specifies the contents of the expression using the ISO 8601 standard. The example in (1) shows the annotation of a fully specified DATE TIMEX3.

(1) a. April 7, 1980
     b. <TIMEX3 tid="t1" type="DATE" value="1980-04-07"
         temporalFunction="false">
         April 7, 1980
         </TIMEX3>

April 7, 1980 is a fully specified temporal expression because it includes all of the information needed to give its value. Many temporal expressions are not fully specified and require additional information from other temporal expressions to provide their full value. We will say more about the annotation of these expressions shortly, but, for now, notice that the annotation in (1) includes an attribute called temporalFunction and that it is set to "false". When a temporal expression requires more information to complete its annotation, this attribute is set to "true" to indicate that a temporal function will be used. For more on this process, refer to the section below on temporal functions.

While the DATE type is used to annotate most calendar expressions, the TIME type is used to capture expressions whose granularity is smaller than one day. Examples of this include *4:20* and *this morning*. Example (2) shows the annotation of a fully specified TIME TIMEX3. Notice that for a TIME to be fully specified, it must include date information as well.

(2) a. 10:30am April 7, 1980
     b. `<TIMEX3 tid="t1" type="TIME" value="1980-04-07T10:30"`
         `temporalFunction="false">`
         `10:30am April 7, 1980`
         `</TIMEX3>`

Expressions such as *for three months* include a DURATION TIMEX3. The **value** attribute of a DURATION again follows the ISO 8601 standard. For example, *three months* receives a **value** of "P3M". Occasionally, a DURATION will appear anchored to another temporal expression. Since TimeML strives to annotate as much temporal information as possible, this information is also included in the annotation of a DURATION with the **beginPoint** and **endPoint** attributes as shown in (3).

(3) a. two weeks from December 17, 2005
     b. `<TIMEX3 tid="t1" type="DURATION" value="P2W"`
         `beginPoint="t2" endPoint="t3">two weeks</TIMEX3>`
     c. `<TIMEX3 tid="t2" type="DATE" value="2005-12-17">`
         `December 17, 2005</TIMEX3>`
     d. `<TIMEX3="t3" type="DATE" value="2005-12-31"`
         `temporalFunction="TRUE" anchorTimeID="t1"/>`

The example in (3a) contains to temporal expressions separated by a signal (see subsection 2.3). The first, *two weeks*, is annotated as a DURATION. The second, *December 17, 2005*, is a fully specified DATE. Every TIMEX3 annotation includes an identification number. This number is used to relate the temporal expression to other TimeML objects. In this case, the identification value in (3c), "t2", is included in the annotation of *two weeks* as the **beginPoint** of the duration. With this information, the **endPoint** of the duration can be calculated. An additional TIMEX3 is created to hold its value. This is the TIMEX3 given in (3d). Since the value of the new TIMEX3 must be calculated, **temporalFunction** is set to "true" and a temporal anchor is supped. This new attribute will be explained below.

The final type of TIMEX3 is used to capture regularly recurring temporal expressions such as *every three days*. This type, SET, uses the attributes **quant** and **freq** to annotated quantifiers in an expression and the frequency of the expression, respectively. An example is given in (4).

(4) a. two days every week
     b. `<TIMEX3 tid="t1" type="SET" value="P2D" quant="EVERY"`
         `freq="1W">two days every week</TIMEX3>`

Temporal Functions. When a temporal expression is not fully specified, it requires the use of a temporal function to calculate its value. In a manual annotation, the user provides a particular anchor time ID that supplies the missing information. The user then gives the correctly calculated value for the `TIMEX3`. In automatic annotation, a library of temporal functions is used to perform the calculation.

The example in (3d) shows an annotation that uses a temporal function. In this case, the end point of a duration was calculated using the `beginPoint` and `value` of the duration given in (3b). For the new temporal expression in (3d), the `temporalFunction` attribute is set to "TRUE" and the `tid` for the duration is given as the `anchorTimeID`. Finally, the correct `value` is supplied. This same process is used for temporal expressions that are missing information such as *April 7*, which is missing the year, and for relative temporal expressions such as *today*.

2.2   Events

Events that can be anchored or ordered in time are captured with TimeML. Such events are predominantly verbs, but nouns, adjectives, and even some prepositions can also be eventive. TimeML events are annotated with the `EVENT` tag. This tag has three main attributes: an ID number, an event instance ID number, and an event class. The classification of an event can help determine what relationships that event may participate in. For example, an event classified as REPORTING will be the first element of an evidential `SLINK` (see the subsection on Subordinating Links in section 2.4). There are seven event classes:

- REPORTING: *say, report, tell*
- PERCEPTION: *see, watch, hear*
- ASPECTUAL: *initiate, terminate, continue*
- I_ACTION: *try, investigate, promise*
- I_STATE: *believe, want, worry*
- STATE: *on board, live, seek*
- OCCURRENCE: *land, eruption, arrive*

Several of these classes introduce an event argument and are of particular interest to the work in this paper. The TimeML Annotation Guidelines [4] detail exactly which events fall into which classes.

Besides the classification of an event, natural language documents supply much more information about events that we need to represent in an accurate annotation. In addition to the head of the event that is captured in the text, an event may include further tense and aspect indicators or modifiers that affect its modality or polarity. Therefore, along with the ID numbers and event class, tense, aspect, part of speech, modality, and polarity information are also stored in the `EVENT` tag. Earlier specifications of TimeML put this information in a separate tag called `MAKEINSTANCE`. The motivation for this tag was to account for multiple instances of a single event. However, to simplify the annotation, this tag has been removed and the information stored in it is now stored with the event itself.

In some cases, a single mention of an event in the text can actually refer to multiple instances, as in example (5).

(5) John swims on Monday and Tuesday.

Here, there is one mention of *swim* that is tagged as an OCCURRENCE EVENT. The TimeML annotation should link this event to the temporal expressions also present in the sentence. However, it is clear that the *swim* event that takes place on *Monday* is not the same one that takes place on *Tuesday*. Instead, it is an instance of the event that is anchored to each temporal expression. With the removal of the MAKEINSTANCE tag, this is accomplished by adding a non-consuming EVENT that holds the correct instance information.

Instances of events can also have different tense, aspect, polarity, or modality properties. When an additional instance of an event is needed, a non-consuming EVENT tag is created to hold information on that instance. For the sake of consistency, the event instance ID number that is included with all EVENTs, even if only one instance is needed, is used to show that an instance of an event participates in a TimeML relationship as seen in section 2.4.

2.3 Signals

When temporal objects are related to each other, there is often an additional word present whose function is to specify the nature of that relationship. These words are captured with the SIGNAL tag, which has one attribute that provides an identification number. Example (6) shows a typical use of preposition *at* as SIGNAL, and a complete annotation of all the temporal objects present.

(6) a. The bus departs at 3:10 pm.
　　 b. The bus
```
<EVENT eid="e1" eiid="ei1" class="OCCURRENCE"
pos="VERB" tense="PRESENT" aspect="NONE" polarity="POS">
departs
</EVENT>
<SIGNAL sid="s1">
at
</SIGNAL>
<TIMEX3 tid="t1" type="TIME" value="XXXX-XX-XXT15:10">
3:10pm
</TIMEX3>
```

2.4 Links

TimeML uses three varieties of LINK tag to represent relationships among temporal objects. In all cases, the LINK tag is non-consuming as there may not be any explicit text to capture or the relationship could be between objects whose locations vary greatly. Each link tag comes with a set of relation types to specify the nature of the relationship. In the following paragraphs, we briefly describe each of these tags: TLINK, ALINK, and SLINK.

Temporal Relationships. All temporal relationships are represented with the TLINK tag. TLINK can be used to annotate relationships between times, between events, or between times and events. In this way, TimeML can both anchor and order temporal objects. A `signalID` can also be used in a TLINK if it helps to define the relationship. The TLINK in example (7) completes the annotation of *The bus departs at 3:10pm.*

(7) `<TLINK lid="l1" eventInstanceID="ei1" relatedToTime="t1"`
     `signalID="s1" relType="IS_INCLUDED"/>`

The possible `relType` values for a TLINK are based on Allen's thirteen relations [5].[1] The relationships for TLINK include before and after, immediately before and after, included, during, simultaneous, and begins and ends. TLINK is also used to assert that two event instances refer to the same event using the IDENTITY `relType`.

Aspectual Links. Events classified as ASPECTUAL introduce an ALINK. These include events such as *begin, stop* and *continue.* The ALINK represents the relationship between an aspectual event and its argument event. In some ways, the ALINK is like a combination of a TLINK and an SLINK as it indicates both a relationship between two temporal elements and aspectual subordination. The possible relationship types for ALINK are: initiates, culminates, terminates, continues, and reinititiates.

Subordinating Links. As mentioned in section 2.2, certain event classes introduce a subordinated event argument. Some examples are verbs like *claim, suggest, promise, offer, avoid, try, delay, think*; nouns like *promise, hope, love, request*; and adjectives such as *ready, eager, able, afraid.* In the following sentences, the events selecting for an argument of situation or proposition type appear in bold face, whereas the corresponding argument is underlined:

(8) a. The Human Rights Committee **regretted** that discrimination against women persisted in practice.
     b. Uri Lubrani also **suggested** Israel was **willing** to withdraw from southern Lebanon.
     c. Kidnappers **kept** their **promise** to kill a store owner they took hostage.

In TimeML, subordination relations between two events are represented by means of a Subordinating Links (or SLINKs). The SLINK tag is perhaps the best example of the current treatment of arguments in TimeML. Reference to each

---

[1] See [22] for details on the mapping of these relations into TimeML.

event is expressed by a pointer to them (through the attributes `eventInstan-ceID` and `subordinatedEventInstance`), and the relation type is conveyed by means of the attribute `relType`, which captures the type of modality projected in each case onto the event denoted by the subordinated clause. `relType` can be any of the following types:

1. `FACTIVE`: When the argument event is entailed or presupposed. Here is an annotated example:[2]

   (9) a. The Human Rights Commitee regretted that discrimination against women persisted in practice.

   b. ```
   The Human Rights Committee
   <EVENT eid="e1" class="I_ACTION">
   regretted
   </EVENT>
   that discrimination against women
   <EVENT eid="e2" class="ASPECTUAL">
   persisted
   </EVENT>
   in practice.
   <SLINK eventInstanceID="e1" subordinatedEventInstance="e2"
   relType="FACTIVE"/>
   ```

2. `COUNTERFACTIVE`: When the main predicate presupposes the non-veracity of its argument:

   (10) a. A Time magazine reporter avoided jail at the last minute...

   b. ```
   A Time magazine reporter
   <EVENT eid="e1" class="I_ACTION">
   avoided
   </EVENT>
   <EVENT eid="e2" class="STATE">
   jail
   </EVENT> at the last minute...
   <SLINK eventInstanceID="e1" subordinatedEventInstance="e2"
   relType="COUNTERFACTIVE"/>
   ```

3. `EVIDENTIAL`: Typically introduced by `REPORTING` or `PERCEPTION` events, such as *tell, say, report* and *see, hear*, respectively.
4. `NEGATIVE_EVIDENTIAL`: Introduced by `REPORTING` and `PERCEPTION` events conveying negative polarity; e.g., *deny*.
5. `MODAL`: For annotating events introducing a reference to possible world.

   (11) a. Uri Lubrani also suggested Israel was willing to withdraw from southern Lebanon.

---

[2] For the sake of simplicity, in this and the following examples we obviate the annotation of the part of speech, tense, aspect, modality, and polarity information in each event and use the eventID number as a reference in the links. In a complete annotation, this information would be included and the event instance ID would be used as a reference in the LINK tags.

b. Uri Lubrani also
```
<EVENT eid="e1" class="I_ACTION">
suggested
</EVENT>
Israel was
<EVENT eid="e2" class="I_STATE">
willing
</EVENT>
to
<EVENT eid="e3" class="OCCURRENCE">
withdraw
</EVENT>
from southern Lebanon.
<SLINK eventInstanceID="e1" subordinatedEventInstance="e2"
relType="MODAL"/>
<SLINK eventInstanceID="e2" subordinatedEventInstance="e3"
relType="MODAL"/>
```

6. CONDITIONAL" For annotating conditional contexts.

(12) a. If Graham leaves today, he will not hear Sabine.

b.
```
<SIGNAL sid="s1"
If
</SIGNAL>
Graham
<EVENT eid="e1" class="OCCURRENCE">
leaves
</EVENT>
<TIMEX3 tid="t1" type="DATE" value="XXXX-XX-XX"
temporalFunction="true" >
today
</TIMEX3>
, he will not
<EVENT eid="e2" class="OCCURRENCE">
hear
</EVENT
Sabine.
<SLINK eventInstanceID="e1" subordinatedEventInstance="e2"
signaled="s1" relType="CONDITIONAL"/>
<TLINK eventInstanceID="ei1" relatedToEventInstance="ei2"
relType="BEFORE"/>
```

The goal of a TimeML annotation is to provide the most complete temporal picture of a text possible. On the surface, it may seem as if temporal links are all that are needed to achieve this. Yet, subordinating and aspectual links are also an essential part of this process. ALINKs contribute information about the internal temporal structure of the events in question. On the other hand, temporal relations can be inferred from some SLINKs. For example, a TLINK of BEFORE relType can be derived between an event and its embedded infinitival clause holding an SLINK of MODAL `relType`. That is, if the subordinated event does happen, it will most likely occur after the subordinating event.[3]

The extraction of new temporal links from subordinating or aspectual relationships should not be mistaken as a replacement for those SLINKs and ALINKs. On the contrary, they still play a vital role in the annotation. The example described above is a case and point: if the new BEFORE TLINK were to replace the SLINK, then there would be no remaining evidence that the subordinated event may or may not have actually happened.

Thus, annotating the relation between proposition- or event-denoting expression and the predicate that selects them contributes two substantial benefits:

a. It enables distinguishing between events that are presented as extensional and those characterized as intensional, a feature that is fundamental for any subsequent task involving temporal ordering and reasoning over events.
b. It expresses a linkage from which additional temporal relations can be automatically derived.

SLINKs, ALINKs, and some TLINKs are examples of how TimeML already accounts for some event arguments. In the next section, we explore this further and consider the impact of adding entity arguments to the annotation.

3   Events and Their Participants

We will assume for our discussion that events can be represented as first order individuals, existentially quantified in a neo-Davidsonian manner where participants to the event are conjoined relations between individuals and the event ([7], [8]). For each event, e, we will identify the participants to this event with a three-place relation, Arg, between arguments e, x (of type individual), and k (of type integer).

(13) $\lambda k\colon \text{int}\lambda x\colon \text{ind}\lambda e\colon \text{event}[Arg(k, e, x)]$

Rather than labeling arguments with specific named semantic functions, such as agent, patient, and instrument, we identify the argument by an index, k. The idea is that a post-parsing procedure will identify the appropriate semantic role played by an argument.

Both named entity arguments and event arguments are expressible in this fashion. For example, for the sentence in (14a), the participants are directly identified by their indices 1 and 2, respectively, but not functionally, as *Agent* and *Patient*.

(14) a. John kissed Mary.
      b. $\exists e[kiss(e) \wedge Arg(1, e, j) \wedge Arg(2, e, m)]$

---

[3] This has been explored within the TARSQI project [6], aimed at creating a toolkit for doing automatic TimeML annotation.

Notice that the current TimeML representation of (14a) identifies the event predicate but not its arguments.

(15) `John`
```
<EVENT eid="e1" eiid="ei1" class="OCCURRENCE"
pos="VERB" tense="PAST" aspect="NONE" polarity="POS">
kissed
</EVENT>
Mary.
```

With the addition into TimeML of an *Arg*-relation, we would be able to identify the entity participants as represented in (14b) above. This should be done cautiously, however, without complicating the specification language or making the annotation task more difficult than it already is. We will take up this issue in Section 5 below.

By design, TimeML treats predicates that select for event arguments differently from those taking named entities. For example, the event-embedding predicate *see*, in most cases, allows the same simple conjunctive representation over arguments that we saw in (14b), assuming the argument is extensional.[4]

(16) a. John saw Mary fall.

  b. $\exists e_1 \exists e_2 [see(e_1) \wedge Arg(1, e_1, j) \wedge Arg(2, e_1, e_2) \wedge fall(e_2) \wedge Arg(1, e_2, m)]$

In the next section, we turn to the question of how to generalize the encoding of an event argument as expressed in TimeML through SLINKs.

3.1   SLINK Encodes Partial Argument Structure

According to the TimeML specification, predicates in natural language that are encoded as introducing SLINKs in fact already identify the embedded complement as an argument to the verb.

For example, the TimeML markup of (17a) explicitly identifies the embedded complement (verb) as a subordinated argument to the event *regret*.

(17) a. John regretted that Sue marrried Bill.

  b. `John`
```
<EVENT eID="e1" class="I_ACTION">
regretted
</EVENT>
that Sue
<EVENT eID="e2" class="OCCURRENCE">
married
</EVENT>
Bill.
<SLINK eventID="e1" subEventID="e2" relType="FACTIVE"/>
```

---

[4] We assume that the typing on the *Arg* relation can be generalized to allow events as arguments.

As it happens, with a factive predicate such as *regret* we can existentially quantify the event representing the embedded complement of the SLINK predicate. A first-order neo-Davidsonian representation of this sentence would, therefore, look like the following:

(18) $\exists e_1 \exists e_2 [regret(e_1) \wedge Arg(1, e_1, j) \wedge Arg(2, e_1, e_2) \wedge marry(e_2) \wedge Arg(1, e_2, s) \wedge Arg(2, e_2, b)]$

The current TimeML representation of this sentence, however, expressed as a first-order expression, is closer to that shown in (19), since no entity arguments are represented in TimeML.

(19) $\exists e_1 \exists e_2 [regret(e_1) \wedge Arg(2, e_1, e_2) \wedge marry(e_2)]$

For all other modality-introducing predicates, TimeML is generally descriptively adequate in differentiating the modal force of the complement expression. For example, the SLINK predicate *believe* is annotated as (20b) below.

(20) a. John believes that Bill went to Japan.
      b. John
```
<EVENT eID="e1" class="I_ACTION">
believes
</EVENT>
that Bill
<EVENT eID="e2" class="OCCURRENCE">
went
</EVENT>
to Japan.
<SLINK eventID="e1" subEventID="e2" relType="MODAL"/>
```

The modal subordination introduced by the propositional attitude predicate *believe* is represented by an SLINK with a `relType` value of MODAL. To model this, we will introduce a special first order variable, $\hat{e}$, effectively encoding the modality of the event and the domain of its subordination. On this strategy, a first order expression representing the partial argument structure of (20b) would be that shown in (21).[5]

(21) $\exists e \exists \hat{e} [believe(e) \wedge Arg(2, e, \hat{e}) \wedge go(\hat{e})]$

We have now partial argument structure for some predicates, but there are many other verbs which select events as their direct arguments as well. This is for example the case of causative predicates (*cause, led*, etc.). The question is, should they be treated in TimeML, and if so, how? The following section will focus on these kind of predicates, proposing a solution to account for them in a way parallel to SLINK relations. We will also show that the same treatment needs to be applied to other relations generally expressed at the discourse level.

---

[5] This is similar to the first order representations in DAML for modal subordination (Jerry Hobb (p.c); cf. http://www.daml.org/ontologies/

4 Encoding Discourse Relations in TimeML

4.1 Relations of Causation

Causative predicates express a specific relation between their event arguments, that of causation, which contributes basic temporal information in a way similar to SLINKs. The representation of causation between event denoting expressions within the same sentence is common in natural languages. For example, the following sentences express causal (and hence temporal) relations between events.

(22) a. [The rain]$_{e1}$ caused [the flooding]$_{e2}$.

   b. [The rioting]$_{e1}$ led to [curfews]$_{e2}$.

   c. [Fifty years of peace]$_{e1}$ brought about [great prosperity]$_{e2}$.

The information provided by this type of relations has been so far ignored in TimeML. However, we believe that there should be an explicit representation of this relation in an event ordering markup language such as this. To capture it, we introduce a new link type, called DLINK for discourse link.

DLINKs will encode the causal relation between two events denoting respectively the cause and the effect, as expressed by verbs like the following, in their causative sense:

(23) *cause, stem from, lead to, breed, engender, hatch, induce, occasion, produce, bring about, produce, secure.*

By means of a DLINK, a sentence such as (24a) can be explicitly annotated as involving a causal relation:

(24) a. The rioting led to curfews on November 22, 2004.

   b. ```
The
<EVENT eid="e1" eiid="ei1" class="OCCURRENCE"
pos="NOUN" tense="NONE" aspect="NONE">
rioting </EVENT>
<EVENT eid="e2" eiid="ei2" class="CAUSE"
pos="VERB" tense="PAST" aspect="NONE">
led </EVENT>
to
<EVENT eid="e3" eiid="ei3" class="OCCURRENCE"
pos="NOUN" tense="NONE" aspect="NONE>
curfews </EVENT>
on
<TIMEX3 tid="t1" type="DATE value="2004-11-22">
November 22, 2004
</TIMEX3>.

<DLINK eventInstanceID="ei1" relatedToEvent="ei3"
relType="CAUSAL" signalID="ei2"/>
<TLINK eventInstanceID="ei3" relatedToTime="t1"
reltype="IS_INCLUDED"/>
```
```

Note that both the subject and object event expressions are syntactically arguments to the causal predicate. In this case, the *Arg* relation is not operative since the matrix predicate is itself a realization of a *Cause* relation directly:

(25) a. The rioting led to curfews.

    b. $\exists e_1 \exists e_2 [rioting(e_1) \land Cause(e_1, e_2) \land curfews(e_2)]$

In addition to causative constructions like those in (22), there are many others where causation is expressed through an explicit causative predicate as well, and yet the relation is not syntactically between two events but between an individual and an event. Consider:

(26) a. $[\text{John}]_x$ caused $[\text{a fire}]_{e2}$.

    b. $[\text{The drug}]_x$ induced $[\text{a seizure}]_{e2}$.

In such cases of event metonymy ([9], [10]), we will introduce a skolemized event instance, $ei1$, to act as the proxy in the causation relation. Hence, the TimeML for (26a) would be as follows below:[6]

(27)
```
John
   <EVENT eid="e1" eiid="ei1" class= "NONE"
   tense="NONE" aspect="NONE"/>
   <EVENT eid="e2" eiid="ei2" class="CAUSE"
   pos="VERB" tense="PAST" aspect="NONE">
   caused </EVENT>
   a
   <EVENT eid="e3" eiid="ei3" class="OCCURRENCE"
   tense="NONE" aspect="NONE">
   fire </EVENT>

   <DLINK eventInstanceID="ei1" relatedToEvent="ei3"
   relType="CAUSAL" signalID="ei2"/>
```

In English as in most languages, causation can be expressed as a relation between elements within the same sentence by means of lexical items such as those presented in (23). But it can also be expressed as a relation between events in two different sentences, connected by causative markers such as *because, since,* or *given that.* Consider:

(28) *Because* [the drought reduced U.S. stockpiles]$_{s1}$, [they have more than enough storage space for their new crop]$_{s2}$.

The same type of DLINKs can be used to express the causation here:[7]

---

[6] Note that the interpretation of *John* as the agent of an event involved in the causation is out of the scope of TimeML; it would be the responsibility of subsequent semantic interpretation to bind the entity *John* to the causing event.

[7] Note that the relation between the events expressed as *drought* and *reduced* is also of causal nature. In this case, it is entailed by the predicate *reduce*, which denotes change of state as part of its core meaning. TimeML aims at being as much surface-based as possible, and so we will not consider cases like this one for the moment.

```
(29) <SIGNAL sid="s1">
     because </SIGNAL>
     the
     <EVENT eid="e1" eiid="ei1" class= "OCCURRENCE"
     tense="NONE" aspect="NONE">
     drought </EVENT>
     reduced U.S. stockpiles, they
     <EVENT eid="e2" eiid="ei2" class="STATE"
     pos="VERB" tense="PRESENT" aspect="NONE">
     have </EVENT>
     more than enough food storage space ...

     <DLINK eventInstanceID="ei1" relatedToEvent="ei2"
     relType="CAUSAL" signalID="s1"/>
```

So far, we have proposed introducing a new TimeML entity, DLINK, in order to encode causation relations between two events. What is, however, the main asset of this move in the context of annotating temporal information in text? As already mentioned, causation relations inherently convey basic temporal information which is of relevance for subsequent temporal reasoning: the cause event (at least) starts at a point in time prior to the beginning of the resulting event.

Other discourse relations also entail a temporal component as part of their meaning. Let's see what they are and how they can be accounted for within TimeML.

4.2   Discourse Relations and Temporal Information

There are a number of discourse relation classifications available in the field (e.g., [11], [12], [13], [14], [15], [16]; see [17] for a comparison of some of them). Our analysis here is based on the classification presented in [18], which annotates the Discourse Graphbank corpus [19], because this corpus has recently been used in the TimeML framework on research devoted to the automatic identification of discourse relations [20].

The classification put forward by Wolf et al. define four broad classes of discourse (or coherence) relations, which in some cases split into more specialized subclasses. Table 1 presents them.[8]

A first glance at this table already reveals a two-fold classification between temporally relevant versus non-relevant discourse relations. Of those with temporal consequences there are the classes of Temporal Sequence and Cause-effect relations. Of the other kind, there are Attribution and Resemblance relations.

Classifying Temporal Sequence relations as belonging to the class of those that intrinsically hold temporal information is obvious.[9] As a matter of fact, the information that in GraphBank is encoded by means of these relations is, in TimeML, expressed through temporal links (TLINKs).

---

[8] The definitions and examples are mainly extracted from [18] and [19].

[9] Temporal Sequence corresponds to what other classifications call Narrative.

Table 1. Coherence relations [18]

Resemblance:	Establishing commonalities and contrasts between the discourse segments.
Parallel	e.g., *[John organized rallies for Clinton], and [Fred distributed pamphlets for him].*
Contrast	e.g., *[John supported Clinton], but [Mary opposed him].*
Example	e.g., *[Young aspiring politicians often support their party's presidential candidate]. For instance, [John campaigned hard for Clinton].*
Generalization	e.g., *[John campaigned hard for Clinton in 1992]. [Young aspiring politicians often support their party's presidential candidate].*
Elaboration	e.g., *[A Young aspiring politician was arrested in Texas today]. [John Smith, 34, was nabbed in a Houston law firm while attempting to embezzle funds for his campaign].*
Cause-effect:	Establishing a causal inference between the discourse segments.
Explanation	Understood as the the standard cause-effect relation. e.g., *[John organized rallies for Clinton], and [Fred distributed leaflets].*
Violated expectation	Normally there is a causal relation between the two segments, but here that causal relation is absent. e.g., *[The weather was nice] [but our flight got delayed].*
Condition	The event described in the main clause can only take place if the event described in the *if*-clause also takes place. e.g., *If the system works, everyone will be happy.*
Temporal Sequence:	Defining a temporal sequence between the two discourse units. No causal relation is involved between them. e.g., *[John bought a book], then [he bought groceries].).*
Attribution:	Establishing an evidential relation of reporting type. e.g., *[John said] that [Mary had brought some wine].*

Cause-effect relations as defined in [18] are temporally relevant as well. In the previous subsection, we already argued this for the cases that are characterized here as Explanation relations –that is, the standard cause-effect relation; e.g, (28). The cause event starts (and possibly also ends) before the beginning point of the effect event.

That same temporal inference can be derived from the other two subclasses as well: Violated expectation and Condition. In the former, the event that creates the expectation (equivalent to the cause in Explanation) starts before the event that violates the expectation (equivalent to the result). In the latter, the event described by the *if*-clause will precede the event in the main clause, if that event holds in the world. Interestingly, conditional constructions are already annotated in TimeML by means of (non-lexically triggered) SLINKs [4].

On the other hand, neither Attribution nor any of the Resemblance relations entail specific temporal relations. Note that the Attribution relations in fact encode some of the relations that in TimeML are expressed by means of SLINKs, particularly, SLINKs of EVIDENTIAL `relType` –those expressing some sort of reporting speech act, as in *John said..., according to...*, etc. This is not to say that it is impossible to infer any temporal relationship from this type of SLINK, but the type of TLINK would not be entailed by virtue of the EVIDENTIAL `relType` in the SLINK. Instead, we would have to examine the tense and aspect of the link participants to try to infer a TLINK.

Finally, temporal relations are not entailed in Resemblance relations either, except for the case of Elaboration. The temporal relation between the two events at play in Parallel and Contrast relations can be of any type: one preceding the other, one overlapping with the other, one after the other, etc. As for Example and Generalization, they express a relation between a generic and a particular situation, but do not involve any particular temporal relation.

Elaboration is however the only coherence relation in this class that intrinsically conveys a particular temporal relation, since the elaborating segment details the event or situation expressed in the elaborated one. Hence, a general temporal relation of inclusion can be inferred.

To sum it up, there are four different discourse relations that are of interest to an annotation scheme devoted to the encoding of temporal information in text, such as TimeML:

- **Temporal Sequence**, already captured by TLINKs.
- **Explanation** (our standard cause-effect relation). It encodes a relation of temporal ordering between the events in the two segments. A new link type has already been proposed for annotating it; namely, DLINK.
- **Violated expectation.** Encoding the same temporal relation as Explanation.
- **Condition.** Encoding the same temporal relation as Explanation. TimeML encodes them by means of SLINKs.
- **Elaboration.** Entailing a relation of overlapping of the two events at play.

4.3   DLINK Scheme

Earlier in the section we introduced a new link type in order to annotate causal relations which, when denoted by a causative predicate, are comparable to SLINKs both syntactically but also in terms of the temporal information they contribute to the discourse. We showed that the link type can be extended also to annotating causation relations expressed by means of discourse markers.

Given the analysis presented in the previous subsection, we now propose to generalize the new link in order to include other temporally relevant discourse relations as well. In particular, it will now account for the relations of: elaboration, explanation (i.e., cause-effect), violated expectation, and condition. Note that in earlier versions of TimeML this last relation was previously accounted for by means of SLINKs.

The specification scheme for the new DLINK entity is shown below:

```
<DLINK>
attributes ::= [lid] [origin] eventInstanceID signalID
               subordinatedEventInstance relType
lid ::= ID
{lid ::= LinkID
LinkID ::= l<integer>}
origin ::= CDATA
eventInstanceID ::= IDREF
```

```
{eventInstanceID ::= EventInstanceID}
subordinatedEventInstance ::= IDREF
{subordinatedEventInstance ::= EventInstanceID}
signalID ::= IDREF
{signalID ::= SignalID | EventInstanceID}
relType ::= 'CAUSAL'|'COND'|'ELAB'
```

The different types of relations will be identified through the attribute
relType. The CAUSAL value clusters together the relations of Explanation and
Violated Expectations since they are temporally equivalent. Condition relations,
on the other hand, will be distinguished by means of a different value: that of
COND. The two events there hold the same temporal relation than those in
Explanation and Violated Expectation relations. By contrast, however, they are
marked as intensional. Indeed, representing conditional constructions by means
of DLINKs changes the spec for SLINKs as well, which will not have the COND
relType value anymore. Finally, Elaboration relations, from which a different
TLINK can be derived, will be represented by means of the value ELAB.

5   Binding Entity Arguments in TimeML

In this section, we propose an extension to the current specification of TimeML
to accommodate the treatment of entity arguments. Our goal is to avoid any explicit mention of entities within the TimeML markup. There are two reasons for
this move: first, entity arguments are not temporally sensitive text extents, unlike
event-denoting predicates and temporal expressions; secondly, we wish to avoid
complicating the specification and subsequent annotation task for human or machine tagging. Therefore, our strategy will be to accomplish the argument binding
independent of the event tag itself. Currently, the EVENT tag is defined as follows:

```
<Event>
    attributes ::= eid eiid class
                   tense aspect pos modality polarity
    eid ::= ID
    {eid ::= EventID
    EventID ::= e<integer>}
    eiid ::= ID
    {eiid ::= EventInstanceID
    EventInstanceID ::= ei<integer>}
    class ::= 'OCCURRENCE' | 'PERCEPTION' | 'REPORTING'
              'ASPECTUAL' | 'STATE' | 'I_STATE' | 'I_ACTION'
    tense ::= 'FUTURE'|'PAST'|'PRESENT'|'INFINITIVE'|
              'PRESPART'|'PASTPART'|'NONE'
    aspect ::= 'PROGRESSIVE'|'PERFECTIVE'|
               'PERFECTIVE_PROGRESSIVE'|'NONE'
    pos ::= 'ADJECTIVE'|'NOUN'|'VERB'|'PREPOSITION'|'OTHER'
    modality ::= CDATA
    polarity ::= 'NEG'|'POS'
```

On our approach, this need not change. Rather than add an argument list to
the event —similar to the SUBCAT list in HPSG [21]— we will treat the binding
of particpants to events in a parallel fashion to the treatment of event ordering;
by introducing a new linking relation, called ARGLINK. This will encode, in
TimeML, the binding accomplished by the *Arg* relation defined in (13) above.

```
<ARGLINK>
attributes ::= alid [origin] eventInstanceID ArgID
alid ::= ID
{alid ::= ArgLinkID
ArgLinkID ::= al<integer>}
origin ::= CDATA
eventInstanceID ::= IDREF
{eventInstanceID ::= EventInstanceID}
ArgID ::= IDREF
{ArgID ::= EntityID}
```

Now let us see how the two participants in sentence (30),

(30) John kissed Mary.

can be represented, using the ARGLINK tag. Recall that the desired logical form
for this sentence is:

(31) $\exists e[kiss(e) \wedge Arg(1, e, j) \wedge Arg(2, e, m)]$

Assuming that the named entities in (30) have been identified and indexed, we
can express the bindings shown in (31) as the two ARGLINKs below:

```
(32) John (ai1)
     <EVENT eid="e1" eiid="ei1" class="OCCURRENCE"
     pos="VERB" tense="PAST" aspect="NONE" polarity="POS">
     kissed </EVENT>
     Mary (ai2).
     <ARGLINK alid="al1" eventInstanceID="ei1" ArgID ="ai1"/>
     <ARGLINK alid="al2" eventInstanceID="ei1" ArgID ="ai2"/>
```

This allows us to take advantage of entity tagging information from other resources, while binding these values to the events identified and marked up within
TimeML.

5.1   Event-Based Entity Chronicles

Once a document has been completely annotated in TimeML, we have a very
good idea of what happened when within the document. With the addition of
entity arguments, we also know who participated in these events. What can we
do with all this information? In this section, we present one application that is
in development that should aid in the extraction of information from a TimeML-
annotated text.

Event-based chronicles are designed to track the event-based behavior of an entity over a document collection. This is an important departure from other work on TimeML for two reasons. First, it relies heavily on knowing who the entity participants of an event are. Second, it is cross document, meaning it takes advantage of multiple documents that have been individually annotated with TimeML to do inference over the entire collection.

The basic functionality of these chronicles is to allow a user to only see those events that a particular entity or group of entities are involved in. This will greatly reduce the number of temporal relationships a user has to sort through. By narrowing the focus of the annotation, a user can quickly assertain just what events the entities in question participated in and how those events relate to other events in the document. Moreover, this application could display events from other documents that involve these entities without overwhelming the user with too much information on irrelevant events and their relationships.

This sort of application would not be possible if not for two things. First, it requires as complete a temporal annotation as possible. This means that simply including temporal links would be insufficient since subordinating, aspectual, and discourse links also supply vital temporal information to the annotation. Second, some awareness of entity arguments is obviously needed so that the focus of the annotations can be narrowed. While these entity arguments are not temporal in nature, they are an invaluable part of using the temporal annotation for this kind of application or question answering in general.

6   Conclusions

In this paper, we discussed the role of arguments in an event annotation spec-ification language. We first described how TimeML handles event arguments in subordinating and aspectual contexts, where SLINKs and ALINKs create event-event relations between a predicate and an event-denoting argument. We proposed that TimeML be enriched slightly to include causal predicates, such as *lead to*, since these also involve event-event relations. Finally, we introduced a linking mechanism that allows entities to be identified with the event they participate in, while not including named entity tagging as part of TimeML.

References

1. Mani, I., Wilson, G.: Robust temporal processing of news. In: ACL 2000. Proceed-ings of the 38th Annual Meeting of the Association for Computational Linguistics, New Brunswick, New Jersey, pp. 69–76 (2000)
2. Schilder, F., Habel, C.: From Temporal Expressions To Temporal Information: Semantic Tagging Of News Messages. In: ACL-EACL-2001, Toulose, France, pp. 65–72 (July 2001)
3. Ferro, L., Mani, I., Sundheim, B., Wilson, G.: Tides temporal annotation guide-lines. Technical Report Version 1.0.2, MITRE Technical Report (2001) MTR 01W0000041

4. Saurí, R., Littman, J., Knippen, R., Gaizauskas, R., Setzer, A., Pustejovsky, J.: TimeML Annotation Guidelines (2005), http://www.timeml.org
5. Allen, J.: Towards a general theory of action and time. Artificial Intelligence 23, 123–154 (1984)
6. Verhagen, M., Mani, I., Saurí, R., Knippen, R., Littman, J., Pustejovsky, J.: Automating temporal annotation within TARSQI. In: Proceedings of the ACL 2005 (2005)
7. Davidson, D.: The logical form of action sentences. In: The Logic of Decision and Action (1967)
8. Parsons, T.: Events in the Semantics of English. MIT Press, Cambridge (1990)
9. Pustejovsky, J.: Current issues in computational lexical semantics. In: ACL 1989, pp. xvii–xxv (1989)
10. Pustejovsky, J.: The Generative Lexicon. MIT Press, Cambridge (1995)
11. Hobbs, J.: On the coherence and structure of discourse (1985)
12. Grosz, B., Sidner, C.: Attention, intentions, and the structure of discourse. Journal of Computational Linguistics 12(3), 175–204 (1986)
13. Mann, W., Thompson, S.: Rhetorical structure theory: Toward a funcitonal theory of text organization. Text 8(3), 243–281 (also available at USC/Information Sciences Institute Research Report RR-87-190) (1988)
14. Polanyi, L.: A formal model of the structure of discourse. Journal of Pragmatics 12, 601–638 (1985)
15. Kehler, A.: Coherence, Reference, and the Theory of Grammar. CSLI Publications (2002)
16. Asher, N., Lascarides, A.: Logics of Conversation. Cambridge University Press, Cambridge (2003)
17. Hovy, E.H., Maeir, E.: Parsimonious or profligate: How many and which discourse structure relations? (1995)
18. Wolf, F., Gibson, E., Fisher, A.: meredith Knight: A procedure for collecting a database of texts annotated with coherence relations (2003)
19. Wolf, F., Gibson, E.: Representing discourse coherence: A corpus-based analysis. Computational Linguistics 31(2), 249–287 (2005)
20. Wellner, B., Pustejovsky, J., Havasi, C., Rumshisky, A., Saurí, R.: Classification of Discourse Coherence Relations: An Exploratory Study using Multiple Knowledge Sources. In: 7th SIGDIAL Workshop on Discourse and Dialogue, Sydney, Australia, pp. 117–125 (July 2006)
21. Pollard, C., Sag, I.: Head-Driven Phrase Structure Grammar. CSLI, Stanford, CA (1994)
22. Verhagen, M.: Times Between the Lines, PhD thesis. Brandeis University, Waltham, USA (2004)

Chronoscopes: A Theory of Underspecified Temporal Representations

Inderjeet Mani

The MITRE Corporation
202 Burlington Road
Bedford, MA 01730, USA
imani@mitre.org

Abstract. Representation and reasoning about time and events is a fundamental aspect of our cognitive abilities and intrinsic to our construal of the structure of our personal and historical lives and recall of past experiences. These capabilities also underlie our understanding of narrative language. This paper describes an abstract device called a Chronoscope, that allows a temporal representation (a set of events and their temporal relations) to be viewed based on temporal abstractions. The temporal representation is augmented with abstract events called *episodes* that stand for discourse segments. The temporal abstractions allow one to collapse temporal relations, or view the representation at different time granularities (hour, day, month, year, etc.), with corresponding changes in event characterization and temporal relations at those granularities. The paper situates Chronoscopes in terms of systems for automatically extracting the temporal structure of narratives.

Keywords: abstraction, temporal information extraction, temporal reasoning, granularity.

1 Introduction

Regularities in our experience, in the form of periodic/cyclic events, provide a basis for our systems for time reckoning, and are crucial in the scheduling of activities for a culture as well as a species. Representation and reasoning about time and events are, of course, intrinsic to our construal of the structure of our personal and historical lives and recall of past experiences. In all these different areas of existence, experiences in the world give rise to various inferences about events and their temporal organization.

Psychological research has shed some light on this inferential process. Early work on visual perception [1] suggested that in inferring whether two event stimuli are simultaneous or successive, the inter-stimulus interval is the determining factor. Later, experiments by [2] revealed that this inference is dependent on stimulus duration and persistence of the image in iconic short term memory. Research on consciousness [3] indicated that ordering of sensations is influenced by latencies in stimulus signal propagation and the brain's tendency to adjust for this delay. This can result in sensations being reported as occurring in a different "subjective" order from their "objective" temporal order. In research on auditory perception, [4] found that

F. Schilder et al. (Eds.): Reasoning about Time and Events, LNAI 4795, pp. 127–139, 2007.

estimates of temporal durations depend on prior expectations and particular modes of attending. Further work on durations [5] showed that the structure of auditory events influences retrospective judgments of their duration.

Natural language has a variety of devices to communicate information about events and their temporal organization, including the use of finite verbs and event nouns, tense and aspect markers, and temporal adverbials, with these devices playing different roles in different languages. While such linguistic information is often present in natural language discourses, they are often vague as to the precise temporal relationships. In (1), from [6], while we know that John arrived at Mary's house after Mary left for dinner, we don't know which started earlier, John's hurrying or Mary's leaving, These events, along with their temporal durations, are left underspecified, resulting in a partial ordering.

> (1) John hurried to Mary's house after work. But Mary had already left for dinner.

Further, crucial information about temporal relationships is often not overtly expressed in the text. In (2), temporal adverbials and aspectual information cue the inference that the twisting occurred during the running, while commonsense knowledge suggests that the pushing occurred before the twisting.

> (2) Yesterday Holly was running a marathon when she twisted her ankle. David pushed her.

Finally, natural language may express information at different levels of temporal granularity. Mentions of events which have long durations may be interspersed with mentions of more punctual ones, as in (3), from [7] (my italics):

> (3) In the course of *the two decades* he *spent* in Kabul, Babur *led* four *expeditions* into India. His fifth and final *campaign* was *launched* in October 1525: it had a characteristically light-hearted *beginning*: "We mostly *drank* and had morning draughts on drinking days". Between *marches* Babur and his nobles *wrote* poetry, *collected* obscene jokes, and *gave chase* to the occasional rhinoceros. Despite internal *dissensions* the Lodis *managed* to *field* an army of 100,000 men and 1,000 elephants against Babur's paltry force of 12,000. The armies *met* on *April 20, 1526,* at the historic battlefield of Panipat a few miles north of Delhi.

The way events are perceived in the world have a strong influence on the interpretation of natural language narratives like (1)-(3). Psychological research by Zwaan [8] has shown that in reading narrative passages, there is a default expectation that successive sentences will describe chronologically successive and contiguous (i.e., temporally adjacent) events. Deviations from this narrative format (as we have in our example passages above) will result in delays in processing information. Further research, such as [9], has confirmed and elaborated this finding, supporting the hypothesis that readers build models of the situation described by the narrative, including representations of temporal directions and distances between events. Kelter et al. [10] have found that when processing a narrative consisting of a sequence of immediately successive events without a temporal shift (i.e., a shift that would be marked by a temporal adverbial), readers took longer to access events that, although mentioned recently, were temporally somewhat remote from the current narrative 'now'. They argue that readers construct dynamic models of the situation, simulating the events in the narrative, and as part of that simulation they represent the temporal distances between events. A temporal shift results in a new model being constructed; no temporal distance effect was found when the text had a temporal shift.

Research in computational linguistics further suggests that readers find it difficult to infer fine-grained temporal relations from natural language. A pilot experiment [11] with 8 subjects providing event-ordering judgments on 280 clause pairs revealed that people have difficulty distinguishing whether there are gaps between events. The subjects were asked to distinguish whether an event was (i) strictly before the other, (ii) before and extending into the other, or (iii) simultaneous with it. These distinctions can be hard to make, as in the example of ordering *try on* with respect to *realize* in (4):

(4) a. Shapiro said he <u>tried on</u> the gloves
    b. and <u>realized</u> they would never fit Simpson's larger hands.

Agreement between 3 subjects was relatively weak (.5 Kappa), but became acceptable (.61 Kappa) if the distinction between (i) and (ii) was ignored.

The above body of psychological literature establishes correspondences between formal aspects of the temporal structure of discourse and the mental representations readers construct. Specifically, the order in which events are narrated, their chronology, the durations of and temporal distances between events, and the narrator's explicit shifts in reference times, marked by temporal adverbials, are all important features used in constructing mental models of narratives. Interestingly, these features are of the sort that can be constructed automatically by information extraction systems, e.g., [12]. However, given that there is vagueness, partial ordering, missing information, and different granularities of temporal representation in narrative language, the representations humans construct from reading narratives can at best be only approximate. This paper focuses on a method of folding such approximation into these formal models, based on a model of granularity.

I will describe an approach that allows one to represent and reason about time and events in the face of vagueness, missing information, and different granularities of temporal representation. Such a representation, called a Chronoscope, aims at capturing some of the subjectivity and looseness, as well as the flexibility inherent in our construction of temporal representations from natural language. It is closely tied to recent frameworks for formal reasoning about time [13][14][15]. Unlike some other theories of granularity and abstraction, it also allows for a straightforward implementation and embedding in various temporal reasoning and visualization tools.

2   Representational Distinctions for Natural Languages

2.1   Underspecified Temporal Relations

Consider example (4) again. Subjects found it difficult to distinguish between the event of Shapiro trying on the gloves being (i) entirely before the event of realizing, versus (ii) extending into and overlapping into the event of realizing. We can represent the temporal relations involved in relations (i) or (ii) as a disjunction[1] of the well-known interval relations defined by Allen [13]:

(5) ea [< m o] eb

These disjunctions of Allen relations can be reified as new, coarse-grained relations. In temporal calculi, this notion has been explored in the work of Freksa [15]. In particular, [< m o] corresponds to Freksa's relation ob (older and survived by), which has a semantics of the start (end) of the first interval preceding the start (end) of the second. Freksa defines 16 coarse-grained relations based on reified disjunctions involving the 13 fine-grained relations specified by Allen. Building on Freksa's work, [16] and [17] have described how relations can be arranged in a hierarchy based on relations between start and end points. In Section 3, we will show how the representation of different levels of coarse- and fine-grained relations can be captured by the chronoscope concept of Zooming.

2.2 Abstract Events and Temporal Discourse Structure

In addition to reification of temporal relations, it is possible to reify abstract events corresponding to entire discourses. Consider discourse (6), from [18]:

(6) a. John went into the florist shop.
b. He had promised Mary some flowers.
c. She said she wouldn't forgive him if he forgot.
d. So he picked out three red roses.

It is clear that b and c comprise a sub-discourse in the main discourse. If we represent these discourses by abstract events (i.e., e0 for the root discourse, e1 for the sub-discourse), then individual events can be viewed as being temporally included ($\subseteq$) in their immediate discourse, giving rise to a tree-structured representation of discourse [19]. This is shown in Figure 1.

In this tree representation, the dominance relation between nodes corresponds to temporal inclusion. The tree is unordered in terms of precedence relation, though by convention the nodes are ordered in order of mention. Abstract events representing

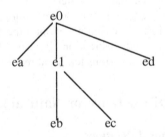

Fig. 1. Episodes in the narrative in (6)

---

[1] The disjunction is represented here using square brackets.

(properly contained) sub-discourses will be called **episodes**. So, the temporal relations in (6) are given by (7), where there is one episode $e1$[2]:

$$(7)\ ea \subseteq e0\ \&\ e1 \subseteq e0\ \&\ ed \subseteq e0\ \&\ eb \subseteq e1\ \&\ ec \subseteq e1\ \&\ ec < ea\ \&\ ea < ed$$

States are represented as minimally included ($\subseteq_{min}$) in the embedding event, without committing to whether the state extends before or after the event. Note that the ordering of eb and ec is left unspecified rather than reified, since the disjunction set $\{<, >\}$ is inconsistent. See [19] for more details, and a comparison to Discourse Representation Theory (DRT) [20].

In some cases, the attachment of sub-discourses will be ambiguous, i.e., there will be more than one tree possible for a given discourse. As an example, consider discourse (8), from [21]:

(8) a. Yesterday, Jack and Sue went to a hardware store
   b. as someone had stolen their lawnmower.
   c. She had seen a man take it and
   d. had chased him down the street, but
   e. he had driven away in a truck.

The narrative (8) could be analyzed as consisting of the events {ea, eb, and e1}, where episode e1 consists of {c, d, e}. Let us now extend discourse (8) with (9):

(9) f. Later, they went to the police station.
   g. The police were not interested in such a minor crime.

Here, ef could be attached under e0 or e1. We can therefore represent the temporal relations for the discourse (8-9) as shown in (10):

$$(10)\quad ea \subseteq e0\ \&\ eb \subseteq e0\ \&\ e1 \subseteq e0\ \&\ ec \subseteq e1\ \&\ ed \subseteq e1\ \&\ ee \subseteq e1\ \&\ eb <$$
$$ea\ \&\ eb < ec\ \&\ ec < ed\ \&\ ed < ee\ \&\ eg \subseteq ef\ \&\ ef \subseteq [e0\ e1]$$

Such an underspecified representation of ambiguity based on factoring out common elements is similar in some respects to underspecified approaches used quite widely in natural language semantics. Representing underspecified temporal structure of this kind, which is common in narratives, is likely to be highly relevant to the temporal models human construct.

3 Chronoscopes

An abstraction allows information to be viewed at different levels of granularity, and is based on research by [22][23][24][25][26]. **Temporal Abstractions** allow one to collapse temporal relations, or to zoom the representation to different time granularities.

Let E be a set of events and R be a set of binary temporal relations on E. A *Temporal Representation* is a relation $T \subseteq E \times E \times R$. A Temporal Representation

---

[2] Here variables are implicitly existentially quantified. We are ignoring the full logical form, focusing instead on the temporal relations alone.

spanning several years could be abstracted at different grain sizes, e.g., time units such as year, month, week, or day. Let $U=<U_1,...,U_n>$ be a sort of Time Units such that U_i *during* U_{i+1} for $1 \leq i < n$, i.e., $U= <..., YYYY\text{-}MM\text{-}DD_1, YYYY\text{-}MM_2, YYYY_3, YY\text{-}DE_4{}^3, YY_5,..>$.

Let us define the **time-granularity** for an event related by a temporal relation r to a time of granularity g:

$$\forall e \; \forall r \; \forall g \; \text{time-granularity}(e, r, g) \equiv \exists t \; \text{calendarTime}(e, r, t) \; \& \qquad \text{(A1)}$$
$$\text{coerceToUnit}(t, U_g)$$

Here, calendarTime(e, r, t) means event e is in the temporal relation r to time t. Also, coerceToUnit(t, U_g) means that time t can be mapped to time unit U at granularity g. For example, "June 1974" could be coerced to "1974", so that we have coerceToUnit(1974-06, U_3); or else it could be coerced to the 1970's, with coerceToUnit(1974-06, U_4), or to the 1900's, with coerceToUnit(1974-06, U_5), etc.

Let's now consider **equi-granular** ($\sim_{rg}$) events:

$$e1 \sim_{rg} e2 \equiv \text{time-granularity}(e1, r, g) \; \& \; \text{time-granularity}(e2, r, g) \qquad \text{(A2)}$$

For example, in (3), *launched* $\sim_{during\text{-}4}$ *met*, i.e., *launched* $\sim_{during\text{-}152X}$ *met*. Events whose temporal locations are in the same calendar decade of the same century will be in a common equivalence class corresponding to the decade position of that century.

The Chronoscope requires that we index temporal relations to a particular level of granularity. In particular, when we introduce a coarse-grained relation, it has to be defined in terms of fine-grained relations at the same level of granularity. Axiom (A3) illustrates how this is done for Freksa's relation ob (older and survived by):

Older-and-survived-by at g:

$$\text{(A3)}$$

$$\forall x \; \forall y \; ob_g(x, y) \equiv \text{before}_g(x, y) \; V \; \text{meets}_g(x, y) \; V \; \text{overlaps}_g(x, y)$$

When we index a temporal relation to a particular level of granularity, certain relations are such that they hold at all higher levels of granularity. For example, in (3), the writing of poetry, collecting of jokes, and chasing of rhinoceros, which occurred during the interval October 1525-April 1526, are simultaneous in the 1520's, and so are simultaneous in the 1500's, etc. This generalization is captured in Axiom (A4):

Upward entailment of simultaneity:

$$\text{(A4)}$$

$$\forall x \; \forall y \; \text{simultaneous}_g(x, y) \supset \text{simultaneous}_{g+1}(x, y)$$

As discussed earlier in connection with axiom (A2), even though *launched* in (3) is before *met*, *launched* is equi-granular with *met* with respect to the 1520s. Thus, *launched* can be viewed as simultaneous with *met* at decade or higher granularities. Axiom (A5), which relates equi-granularity to simultaneity, allows us to make such an inference:

Zoomed-simultaneity at g:

$$\text{(A5)}$$

$$\forall x \; \forall y \; x \sim_{during\text{-}g} y \supset \text{simultaneous}_g(x, y)$$

---

[3] DE stands for "decade" in the time expression annotation scheme in TimeML; see [27].

There will be relations which hold at all levels of granularity. For example, in (3), *launched* and *beginning* are simultaneous at all granularities, since they are synonyms of one another. Axiom (A6) allows one to drop the granularity subscript for a given relation r.

Granularity-invariance of relation r:

$$\forall x \; \forall y \; \text{invariant}(r) \equiv r_g(x, y) \equiv r_{g+1}(x, y)$$

(A6)

For any set of events E, let **characterization**(e, E) be true if e can represent E. In general, a characterization is an abstract event corresponding to the individual correlate of some proper subset of events in E. For example, an abstract event that is the individual correlate of the set of events e unique to a month of a particular year could serve as a characterization of the events E of that month. Or E might be characterized in terms of an abstract event from a background ontology. When a set of events E is characterized by an abstract event e, every temporal relation (or link) from an event y not in E to any event in E has to be 'rewired' from y to e, and every link from any event in E to an event y not in E has to be rewired from e to y. In a characterization of E, temporal links among the events in E disappear from view.

With equi-granularity $\sim_{rg}$ as the equivalence relation on a temporal representation s, we can create a hierarchy of partitions $\pi_{\sim rg}(s)$. Let $Z_r(s) = <\pi_{\sim rg1}(s), .., \pi_{\sim rgn}(s)>$ be a sequence such that $\pi_{\sim rgi}(s) \leq \pi_{\sim rgi+1}(s)$ for $1 \leq i < n$, where $\leq$ is a refinement relation. We call $Z_r(s)$ a **Temporal Zooming**, as it permits zooming to any temporal grain size. Given an ordering of time units, a Zooming Z_r allows us to drill down to views of temporal representations based on fine-grained units as well as roll up to views based on coarse-grained units. We call Z_{rj} a *Zooming to grain j*. Thus, a Zooming to year grain U_3, i.e., Z_{r3}, will not look inside the months (instead, it will use characterizations for them).

Consider text (3) again:

(3) In the course of *the two decades* he *spent* in Kabul, Babur *led* four
    expeditions into India. His fifth and final *campaign* was *launched* in
    October 1525: it had a characteristically light-hearted *beginning*: "We
    mostly *drank* and had morning draughts on drinking days". Between
    marches Babur and his nobles *wrote* poetry, *collected* obscene jokes, and
    gave chase to the occasional rhinoceros. Despite internal *dissensions* the
    Lodis *managed* to *field* an army of 100,000 men and 1,000 elephants
    against Babur's paltry force of 12,000. The armies *met* on *April 20, 1526*,
    at the historic battlefield of Panipat a few miles north of Delhi.

Let us assume that r = [= o ⊆ mi], where mi is 'met by'. The resulting Zooming Z_R for the text (3) is as follows[4]:

---

[4] The square brackets are used, in somewhat overloaded fashion, to indicate the set of events in the partition cell that is being characterized. The calendar time associated with the events in the partition cell is indicated as a subscript.

(11) $Z_{r2}(s)$ (month grain):

/* the least granularity in the text is at day grain, so the first grain size to which we roll up is month grain. */

1526-04: $e_{1526-04-20}$

$Z_{r3}(s)$ (year grain):

/*here, sub-year granularities are rolled up[5]. */

1525: $e_{1525-10}$, march, write, collect, give-chase, dissent, manage, field
1526: $e_{1526-04}$

$Z_{r4}(s)$ (decade grain):
152X: e_{1525}, e_{1526}

$Z_{r5}(s)$ (century grain):
15XX: $[_{150/1X}$ spend, lead, expedition], $[_{152X} e_{1525} e_{1526}]$

where
$e_{1526-04-20} = [_{1526-04-20}$ meet$]$
$e_{1526-04} = [_{1526-04} e_{1526-04-20}]$
$e_{1526} = [_{1526} e_{1526-04}]$
$e_{1525-10} = [_{1525-10}$ launch, campaign, drink$]$
$e_{1525} = [_{1525} e_{1525-10}$ march, write, collect, give-chase, dissent, manage, field$]$

It can be seen from (11) that $\pi_{\sim rgi}(s) \leq \pi_{\sim rgi+1}(s)$ for $1 \leq i < n$. Also, since the characterization of a singleton set allows for picking that element as its representative, the *meet* event can be projected upwards. Finally, if launching, campaigning, and drinking are viewed as the single event of *preparing-for-war* (via an event script, for an example), then (11) would simplify to:

(12)
$Z_{r2}(s)$ (month grain):
1526-04: $_{1526-04-20}$meet

$Z_{r3}(s)$ (year grain):
1525: $_{1525-10}$prepare-for-war, march, write, collect, give-chase, dissent, manage, field
1526: $_{1526-04}$meet

$Z_{r4}(s)$ (decade grain):
152X: e_{1525}, $_{1526}$meet

$Z_{r5}(s)$ (century grain):
15XX: $[_{150/1X}$ spend, lead, expedition], $[_{152X} e_{1525 \, 1526}$meet$]$

---

[5] This is an over-simplification for the sake of readability. The marches take place from October 1525-April 1526.

As we will see later, there are a number of tasks where we may need to compare temporal representations. For example, we may want to compare the narrative of two different histories of a famous (or infamous) military campaign. In order to do so, we can consider their intersection $\cap$, i.e., the elements (events E and temporal relations R) in common. Once we have granularity in the picture, however, we can make the intersection sensitive to the granularity. Thus, we can make the narratives being compared look more similar, in terms of intersection, as we zoom out. The following theorem shows that it does not matter whether we zoom first and then intersect, or vice versa.

Distributivity of zooming over intersection:

(A7)

$$\forall s1 \ \forall s2 \ Z_{rg} (s1 \cap s2) \equiv Z_{rg}(s1) \cap Z_{rg}(s2)$$

Given a temporal representation, it can be filtered in various ways depending on the needs of a particular task. Filters can include time constraints, e.g., events on 9/11 or during and after the Hiroshima attack; the particular episode, e.g., the 'Mary' episode in (6); particular participant, etc. A temporal representation constrained based on event participants is called a **trajectory**. A trajectory is the temporal path taken by an event participant (or sets of them), e.g., the trajectory of Babur during his lifetime or during his last invasion of India. Trajectories can be abstracted and intersected at particular levels of granularity. The intersections of trajectories are particularly interesting in the construction of biographical, historical, and literary narratives.

In a Chronoscope, temporal abstraction, filtering, and zooming can be composed together, allowing for considerable flexibility in dealing with the complexity of the cognitive information space. One would expect that in human comprehension of narratives, the events and their temporal ordering can be indexed by agent, and this is indeed an assumption of psychological research [10][11]. Given this assumption, the models constructed by readers are likely to include intersections of agents' trajectories at different levels of granularity.

4 Embodiment

The TimeML annotation scheme for events and their temporal anchors in text [27] forms a basic underpinning for Chronoscopes. Temporal Representations are directly represented in TimeML as temporal links (TLINKs) among events, and between events and times, corresponding to the calendarTime relation described above. Recent research in information extraction has yielded new machine learning approaches that can automatically generate the temporal relations in TimeML [12].

It is often useful to be able to compare two temporal representations, in order to assess reliability of human annotation of temporal representations or of different accounts of a given course of events, to score a machine temporal representation against a human one, or to merge two different temporal representations of similar information for information extraction or summarization purposes. A simple scorer for temporal links has been developed to compare TLINKs in a pair of documents (which are identical except possibly for TLINKs), before or after axioms for transitive closure have been applied. The scorer can easily be extended to allow for user-defined temporal relation equality predicates, e.g., Freksa's pr = [< m], ob = [< m o]. It could

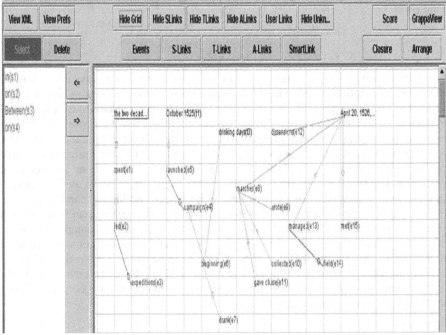

Fig. 2. Graphical Representation of TimeML temporal relations in the narrative in (3)

also be extended to allow for granularity parameter, so that at e.g., year grain, one can successfully match TLINKs with unequal relations, e.g., (launch$_{1525-10}$ < drink$_{1525-10}$) and (launch$_{1525-10}$ > drink$_{1525-10}$).

Chronoscopes can also be embedded in an annotation and visualization environment called TANGO, which is tied to TimeML. Figure 2 shows the graphical display of the TimeML annotation of the example (3). TANGO allows for editing and viewing of TimeML annotations by viewing events and their links in a grid-like display, where times are laid out horizontally and events aligned vertically. TANGO also permits the selection of sets of elements in the display by boxing a region. Built-in temporal filters can be expressed as queries on a temporal database underlying the TimeML representation, and can then be integrated with TANGO by applying those constraints to a boxed region. A list of named entities in a document, attached as arguments to TimeML event predicates, can be used very easily as a filtering mechanism to detect trajectories and their crossings. Episodes require a layer of annotation on top of TimeML, as described in [19]. Once annotated, TANGO can be extended to display episodes by visually marking a set of events and their links that constitute an episode. Zooming can be implemented by sorting the times into bins based on time units, and then constructing different granular representations based on their characterizations.

5 Related Work

This research derives from earlier work by [22], who suggested that in the course of reasoning we conceptualize the world at different levels of granularity, and that in a particular reasoning process we distinguish only those things that are relevant to that process, making other things indistinguishable for all practical purposes. In any given situation, a granularity is determined, allowing the local theory to be selected. When the grain size shifts, certain "articulation axioms" are used to map to another local theory. In general, a mapping which induces a change in granularity can be considered a special case of an *abstraction* in the sense of [23], which lays out a formal theory of abstraction.

Others who have discussed granularity shifts include Euzenat [24], who exploits complex relations which are disjunctions of other relations. Euzenat's approach postulates a number of fundamental algebraic properties of granularity operators. Pianesi and Varzi [25] discuss degrees of temporal granularity in event structure. The analogue of zooming in their account is in terms of a "minimal divisor" on structures corresponding to sets of events, where temporal differences within the divisor are neglected. The question of how granularity shifts may be accommodated within semantic processing of natural language sentences is addressed in [26], which views granularity shifts in terms of abstraction operators on logical forms. Bettini et al. [28] have developed the GSTP system, which allows simple (i. e., non-disjunctive) metric temporal constraints that include granularity, such as "one to three (business) days", to be mapped to different ranges of hours depending on the application. GSTP includes algorithms for temporal constraint satisfaction with multiple granularities.

There have been a variety of systems for temporal visualization of natural language. One particular system that takes granularity into account is that of Matsushita et al. [29], who represent time expressions as points in a plane, with the points connected by lines to express transitions in time. These transition patterns remain similar across scales, and Matsushita et al. argue that they reflect what literary theorists [30] call the "rhythm" of a narrative.

6 Conclusion

The Chronoscope is a simple but flexible device that allows one to represent and manipulate temporal representations inferred from natural language texts. It is tied to existing annotation schemes and ontologies, and can also be integrated with annotation and visualization tools. It is also motivated by psychological considerations.

There are many challenges remaining to be addressed in this work, including the representation of gapped sets of times like the events between marches in (3), generic events like the drinking in (3), and the zooming applied to events which are in the scope of modal operators or subordinated to other events (including the quotation in (3)). While TimeML and OWL-Time have a representation for some of these items, there is much remaining to be done here. Concise characterization in zooming remains a fundamental challenge, however, though work on event summarization and 'event scripts' is clearly relevant. Research on the precision of event durations, and extracting metric as well as qualitative temporal constraints from natural language narratives, as [31] and [32] have explored, are also important to the further automatic extraction of Chronoscopes.

Acknowledgements

I am grateful to two anonymous reviewers for their comments.

References

1. Allport, D.A.: Phenomenal Simultaneity and the Perceptual Moment Hypothesis. British Journal of Psychology 59(4), 395–406 (1968)
2. DiLollo, V.: Temporal integration in visual memory. Journal of Experimental Psychology 109(1), 75–97 (1980)
3. Libet, B., Wright, E., Feinstein Jr., B., Pearl, D.K.: Subjective referral of the timing for a conscious sensor experience: A functional role for the somatosensory specific projection system in man. Brain 194, 191–222 (1979)
4. Jones, M.R.: Dynamic Attending and Responses to Time. Psychological Review 96(3), 459–491 (1989)
5. Boltz, M.G.: Effects of Event Structure on Retrospective Duration Judgments. Perception and Psychophysics 57, 1080–1096 (1995)
6. Dowty, D.: The effects of aspectual class on the temporal structure of discourse: semantics or pragmatics? Linguistics and Philosophy 9, 36–61 (1986)
7. Ghosh, A.: The Man Behind The Mosque, (2005) http://www.amitavghosh.com/
8. Zwaan, R.A.: Processing narrative time shifts. Journal of Experimental Psychology: Learning, Memory, and Cognition 22, 1196–1207 (1996)
9. van der Meer, E., Beyer, R., Heinze, B., Badel, I.: Temporal order relations in language comprehension. Journal of Experimental Psychology: Learning, Memory, and Cognition 28(4), 770–779 (2002)
10. Kelter, S., Kaup, B., Claus, B.: Representing a described sequence of events: A dynamic view of narrative comprehension. Journal of Experimental Psychology: Learning, Memory, and Cognition 30, 451–464 (2004)
11. Mani, I., Schiffman, B.: Temporally Anchoring and Ordering Events in News. In: Pustejovsky, J., Gaizauskas, R. (eds.) Time and Event Recognition in Natural Language, John Benjamins, Amsterdam (2006)
12. Mani, I., Wellner, B., Verhagen, M., Lee, C.M., Pustejovsky, J.: Machine Learning of Temporal Relations. In: Proceedings of the 44th Annual Meeting of the Association for Computational Linguistics (COLING-ACL), Sydney, Australia, pp. 753–760 (2006)

13. Allen, J.: Towards a General Theory of Action and Time. Artificial Intelligence 23, 123–154 (1984)
14. Hobbs, J., Pan, F.: An Ontology of Time for the Semantic Web. ACM Transactions on Asian Language Processing (TALIP): Special issue on Temporal Information Processing 3(1), 66–85 (1984)
15. Freksa, C.: Temporal Reasoning Based on Semi-Intervals. Artificial Intelligence 54(1), 199–227 (1992)
16. Verhagen, M.: Times Between the Lines. Ph.D. thesis. Department of Computer Science. Brandeis University (2005)
17. Schilder, F.: A Hierarchy for Convex Relations. In: Proceedings of the 4th International Workshop on Temporal Representation and Reasoning, pp. 86–93 (1997)
18. Webber, B.: Tense as Discourse Anaphor. Computational Linguistics 14(2), 61–73 (1988)
19. Mani, I., Pustejovsky, J.: Temporal Discourse Models for Narrative Structure. In: ACL Workshop on Discourse Annotation, Barcelona, Spain (2004)
20. Kamp, H., Reyle, U.: Tense and Aspect, pp. 483–546, ch. 5. Kluwer Academic Publishers, Dordrecht (1993)
21. Hwang, C.H., Schubert, L.K: Tense Trees as the fine structure of discourse. In: Proceedings of the 30th Annual Meeting of the ACL, pp. 232–240 (1992)
22. Hobbs, J.: Granularity. In: Proceedings of the International Joint Conference on Artificial Intelligence, pp. 432–435 (1984)
23. Giunchiglia, F., Walsh, T.: A Theory of Abstraction. Artificial Intelligence 57, 2–3 (1992)
24. Euzenat, J.: An Algebraic Approach to Granularity in Qualitative Time and Space Representation. In: Proceedings of IJCAI 1995, pp. 894–900 (1995)
25. Pianesi, F., Varzi, A.C.: Refining Temporal Reference in Event Structures. Notre Dame Journal of Formal Logic 37(1), 71–83 (1996)
26. Mani, I.: A Theory of Granularity and its Application to Problems of Polysemy and Underspecification of Meaning. In: Cohn, A.G., Schubert, L.K., Shapiro, S.C. (eds.) KR 1998. Principles of Knowledge Representation and Reasoning: Proceedings of the Sixth International Conference, pp. 245–255. Morgan Kaufmann, San Francisco (1998)
27. Pustejovsky, J., Ingria, B., Sauri, R., Castano, J., Littman, J., Gaizauskas, R., Setzer, A., Katz, G., Mani, I.: The Specification Language TimeML. In: Mani, I., Pustejovsky, J., Gaizauskas, R. (eds.) The Language of Time: A Reader, Oxford University Press (2005)
28. Bettini, C., Mascetti, V., Pupillo, V., GSTP,: A Temporal Reasoning System supporting Multi-Granularity Temporal Constraints. In: Proceedings of IJCAI 2003, pp. 1633–1634 (2003)
29. Matsushita, M., Ohta, M., Iida, T.: A visualization method of time expressions using starting/ending point plane. In: TIME 1998. Proceedings of Temporal Representation and Reasoning, pp. 162–168 (1998)
30. Genette, G.: Narrative Discourse: An Essay in Method. Cornell University Press (1983)
31. Pan, F., Mulkar, R., Hobbs, J.: Learning Event Durations from Event Descriptions. In: Proceedings of the 44th Annual Meeting of the Association for Computational Linguistics (COLING-ACL), Sydney, Australia, pp. 393–400 (2006)
32. Mani, I., Wellner, B.A: Pilot Study on Acquiring Metric and Temporal Constraints for Events. In: Proceedings of the ACL 2006 Workshop on Annotating and Reasoning about Time and Events (ARTE), Sydney, Australia, pp. 753–760 (2006)

Author Index

Lecture Notes in Artificial Intelligence (LNAI)